ENDOCRINOLOGY RESEARCH AND CLINICAL DEVELOPMENTS

ENDOCRINE DISRUPTING CHEMICALS

OCCURRENCE, EXPOSURES AND HEALTH RISKS

ENDOCRINOLOGY RESEARCH AND CLINICAL DEVELOPMENTS

ENDOCRINOLOGY RESEARCH AND CLINICAL DEVELOPMENTS

ENDOCRINE DISRUPTING CHEMICALS

OCCURRENCE, EXPOSURES AND HEALTH RISKS

SABRINA JOHNSTON
EDITOR

nova publishers
New York

Library of Congress Cataloging-in-Publication Data

Library of Congress Control Number: 2016938500
ISBN: 978-1-63485-231-9

Published by Nova Science Publishers, Inc. † New York

CONTENTS

PREFACE

Wide variety of pesticides and other synthetic chemicals may disrupt the normal function of the endocrine system, cause alteration and also attack the immune system of humans and wildlife at very low concentration levels. Chapter One of this book provides an overview of analytical methods, including sample preparation steps, chromatographic separation and detection of endocrine disrupting pesticides in a variety of food, environmental and biological matrices. Chapter Two studies endocrine disrupting compound (EDC) exposure from recycled wastewater compared to other water related risks in developed and developing nations. Chapter Three summarizes the role of EDCs on pregnancy and birth outcomes, included but not limited to fertility problems, premature birth, low birth weight, and miscarriage along with in-vitro, animal, and epidemiological studies. In the last chapter, the worst EDCs will be revealed and their related potential hazardous health risks to humans are discussed. Furthermore, suggestions to help the reader avoid such toxic chemicals are presented.

Chapter 1 – Wide variety of pesticides and other synthetic chemicals may disrupt the normal function of the endocrine system, cause alteration and also attack the immune system of humans and wildlife at very low concentration levels. This chapter provides an overview of analytical methods, including sample preparation steps, chromatographic separation and detection of endocrine disrupting pesticides in a variety of food, environmental and biological matrices. The major objectives of sample preparation techniques are to isolate the component of interest from the sample; remove potential interferences from the sample matrix; if necessary, increase the concentration of the target analytes at concentration detectable by the subsequent analytical instrumentation. The focus is on the evaluation and improvement of sample preparation methods, among them liquid-based extraction, sorption-based extraction, instrumentally supported extractions and finally, novel modifications of extraction/microextraction procedures are discussed. Current trends include mainly the development of "green" microextraction techniques and the synthesis of new sorbents for sorption-based methods. Special subchapter is devoted to separation methods, where, at present, mainly chromatographic methods are utilized. Modes of mass spectrometric ionization and detection and obtained limits of detection and quantification are reviewed. Special attention is devoted to matrix-effects suppression and elimination. Occurrence of endocrine disrupting pesticides in real-life samples is discussed.

Chapter 2 – Israel is the global leader in wastewater reuse making it the ideal test case to investigate human exposure to contaminants via recycled wastewater pathways. Studies around the world have revealed that wastewater frequently contains endocrine disrupting chemicals (EDCs) that are not completely removed by traditional treatment. As these compounds can harm ecological and human health, it was initially hypothesized that residual EDCs were contributing to reproductive problems in Israel's population. The fate of EDCs from wastewater effluent in Israel and the Palestinian West Bank's shared water basins was assessed over a two year period, and the associated health risks and economic costs of competing wastewater treatment options evaluated. Results from this analysis can inform both developed and developing nations since Israel highly treats and reuses most of its wastewater while only a small percent of sewage undergoes any treatment in the Palestinian territories. EDCs in the region's aquatic environment were measured at concentrations consistent with global averages. According to modeling results, human exposure to bisphenol A via recycled wastewater is likely small compared to ingestion through food. While a potential health risk was identified from bisphenol A in recycled wastewater, it was most pronounced in extreme exposure situations and when reproductive capacity was already compromised. In the West Bank, a cost-effectiveness analysis determined that the marginal improvement in removal rates of estrogens from building tertiary versus secondary level wastewater treatment facilities would not justify the added cost to the Palestinian Authority. Secondary treatment of sewage removed EDCs sufficiently, and the money saved by not investing in tertiary treatment could be better used to provide more essential sanitation services. Tertiary treatment produces other important environmental benefits in which countries like Israel may wish to invest, but it yields only marginally better EDC removal, so developed countries aiming to reduce overall EDC exposure should target sources other than treated wastewater. These findings suggest that EDCs from treated wastewater merit continued research, and demonstrate the importance of environmental cooperation in regions with bifurcated development. However, as EDCs are not the biggest risk to human health in developed or developing nations at this time, they should not be the primary driver for policy governing wastewater management.

Chapter 3 – Endocrine disrupting chemicals (EDC's) are exogenous chemicals that are either plant-derived or man-made, and can alter the functions of the endocrine system and cause various health problems by interfering with the synthesis, metabolism, binding, or cellular responses similar to natural estrogens. Industrialization has introduced a battery of EDC's to the environment which has shown to have drastic outcomes on human health particularly reproduction and pregnancy. Successful pregnancy requires proper orchestration and adaptations of the endocrine system. This article summarizes the role of the EDC's [polybrominated diphenyl ethers (PBDE), 2, 3, 7, 8-tetrachlorodibenzo-p-dioxin (TCDD), bisphenol A (BPA), phthalates (DEHP)s] on pregnancy and birth outcomes, included but not limited to fertility problems, premature birth, low birth weight, and miscarriage along with *in-vitro*, animal, and epidemiological studies. Phytoestrogens are included in the discussion to compare effects of naturally occurring endocrine disrupting chemicals to the man-made ones. Several research demonstrates that fetal exposure to EDC's may negatively affect reproductive health for the lifetime of the organism, increasing chances of reduced fertility, negative pregnancy outcomes, and poor birth outcomes. Epidemiological experiments often associate high concentration of the chemicals with elevated chances of adverse outcomes and reduced fertility. However, there are many missing and conflicting information in this area. Despite uncertainties and knowledge gaps, the available evidence suggests that limiting exposure to EDC's especially during pregnancy is advisable since it can show its affect even two generations latter.

Chapter 4 – In the recent years, or to be exact since the middle of the last decade, the possible adverse effects of synthetic (man-made) and naturally occurring chemicals on the human and animal endocrine system have been the topic of many scientific and medical researches. These concerns were even highlighted further when the research began to show that these chemicals can influence the endocrine system at very low concentrations close to parts per trillion levels where the effects may be cumulative, potentially appearing in subsequent generations. Thus, these resulting effects might be both irreversible and threatening to the human sustainable development. These chemicals, known as Endocrine Disrupting Chemicals (EDCs), can be found in numerous places and products throughout the world since they are used, just to name a few, in everyday consumer products, household items, cleaning products, cosmetics, industrial processes, and waste disposal. EDCs are so widespread since they are heavily and clandestinely utilized by many manufacturers due to their stability, non-degradability, and high profitability. Consequently, humans can be exposed to EDCs through food, water, air, consumer products, and contaminated soil. Once exposed, EDCs can enter the human body via ingestion, inhalation, or skin permeation, where they can reside for a long time and interfere with the body's ability to regulate its growth, metabolism, development, and hormone balance by mimicking, blocking or disrupting the body's natural hormones. A number of observations of adverse health effects have been made in which EDCs could play a role, including but not limited to: declining sperm counts, congenital malformations in children, cancer, retarded sexual development, and retarded neurobehavioral development. In this chapter, the worst EDCs will be revealed and their related potential hazardous health risks to humans will be discussed. Furthermore, suggestions to help the reader avoid such toxic chemicals will be presented.

In: Endocrine Disrupting Chemicals
Editor: Sabrina Johnston

ISBN: 978-1-63485-231-9

Chapter 1

PRESENT-DAY CHALLENGES IN ISOLATION/PRECONCENTRATION AND DETERMINATION OF ENDOCRINE DISRUPTING PESTICIDES BY CHROMATOGRAPHY COMBINED WITH MASS SPECTROMETRY

Svetlana Hrouzková*[*] *and Agneša Páleníková

Institute of Analytical Chemistry,
Faculty of Chemical and Food Technology,
Slovak University of Technology in Bratislava,
Bratislava, Slovak Republic

ABSTRACT

Wide variety of pesticides and other synthetic chemicals may disrupt the normal function of the endocrine system, cause alteration and also attack the immune system of humans and wildlife at very low concentration levels. This chapter provides an overview of analytical methods, including sample preparation steps, chromatographic separation and detection of endocrine disrupting pesticides in a variety of food,

[*]Corresponding author: Institute of Analytical Chemistry, Faculty of Chemical and Food Technology, Slovak University of Technology in Bratislava, Radlinského 9, 812 37 Bratislava, Slovak Republic Email: svetlana.hrouzkova@stuba.sk.

environmental and biological matrices. The major objectives of sample preparation techniques are to isolate the component of interest from the sample; remove potential interferences from the sample matrix; if necessary, increase the concentration of the target analytes at concentration detectable by the subsequent analytical instrumentation. The focus is on the evaluation and improvement of sample preparation methods, among them liquid-based extraction, sorption-based extraction, instrumentally supported extractions and finally, novel modifications of extraction/microextraction procedures are discussed. Current trends include mainly the development of "green" microextraction techniques and the synthesis of new sorbents for sorption-based methods. Special subchapter is devoted to separation methods, where, at present, mainly chromatographic methods are utilized. Modes of mass spectrometric ionization and detection and obtained limits of detection and quantification are reviewed. Special attention is devoted to matrix-effects suppression and elimination. Occurrence of endocrine disrupting pesticides in real-life samples is discussed.

Keywords: endocrine disrupting pesticides, extraction, gas chromatography, instrumental techniques, liquid chromatography, liquid-based extraction, mass spectrometry, matrix effects, microextraction, pesticide residues, sorption-based extraction

INTRODUCTION

Pesticides are a special group of compounds used worldwide for the protection of agricultural crops against pests and diseases. Although the use of these compounds brings benefits to the production in agriculture, many of them reach other than the intended target species and contaminate soil, water and food. Unfortunately, not only pesticides, but also their degradation and transformation products occur in food products or in compartments of the environment, and may be harmful to wildlife and human. Many new reports deal with the problem of metabolites formed during the metabolizms in the organizms. Complex food and environmental samples submitted for analysis represent a challenge to analytical chemists in their efforts to identify and quantify residues of different xenobiotics at trace levels. Intensive concern for the health of society resulted in the strict regulation of maximum residue limits (MRLs) of pesticide residues in food. Various organizations set MRLs, such as the Codex Alimentarius Commission, WHO/FAO (World Health Organization, Food and Agriculture Organization), and European Union (EU) as well as different national governments. Pesticides may disrupt the normal function of the endocrine system and also attack the immune system of humans and wildlife (Colborn et al., 1993). These exogenous substances that interfere with the endocrine system and disrupt the physiologic function of hormones are called endocrine disrupting chemicals (EDCs).

The increased world-wide concern about EDCs has led to a need to develop advanced sensitive and specific analytical tools for their quantification in food, environmental and biological samples at very low concentrations (Hrouzková and Matisová, 2012). Such low concentration levels need multistep sample pretreatment, extraction, pre-concentration, and cleaning.

The state-of-the-art EDCs screening methods, based on biological assays and chromatographic/hyphenated techniques, are limited in their effectiveness (Almeida et al., 2007). For reliable identification and quantification of compounds at ultratrace level methods with chromatographic separation and advanced detection are required. Gas chromatography-mass spectrometry (GC-MS) and GC-tandem MS are used for volatile or volatilizable analytes, while liquid chromatography-tandem mass spectrometry (LC-MS/MS) is aimed at determining more polar and less volatile compounds (Comerton et al., 2009).

This chapter is devoted to pesticides that exhibit or are supposed to exhibit endocrine disrupting properties. Sample preparation and chromatographic analytical methods for the identification and quantification of endocrine disrupting pesticides (EDPs) at ultratrace concentration level are discussed. Modern approaches to the development of a method applicable to a wide variety of compounds with single/combined conventional and advanced instrumental extraction techniques in various matrices or a solventless extraction technique at microscale level are highlighted. Selective MS detection employing current ionization techniques and the most frequently used MS analyzers in detection are covered. Matrix effects present one of today's most challenging analytical issues. A thorough understanding of matrix effects would yield fundamental insights for different food matrices, corresponding sample preparation, and subsequent instrument performance, thus allowing identification and quantitation to be addressed. Occurrence of EDPs found in real-life samples is covered in the individual sub-chapter.

Table 1. Selection of endocrine disrupting pesticides in classification according to EC (European workshop, 1996)

Type of pesticides	Use	Substances – category 1	Substances – category 2
Organochlorines	Herbicides	Acetochlor Alachlor Linuron	Dieldrin Aldrin Diuron Nitrofen 2,4-Dichlorophenol
	Insecticides	Chlordane DDT and metabolites Hexachlorocyclohexane Mirex Chlordecone (Kepone)	Endosulfan (alfa, beta and sulfate) Heptachlor
	Fungicides	Hexachlorobenzenes Vinclozolin	Iprodione Prochloraz Triadimefon
	Acaricide		Dicofol
Organophosphates	Insecticides/ Acaricides	Fenitrothion Omethoate	Acephate Diazinon Dimethoate Chlorfenvinphos Malathion Methylparathion Parathion (Parathion-ethyl) Phosphamidon Trichlorfon (Dipterex)
Triazines	Herbicides	Atrazine Metribuzin Terbutryn	Cyanazine Prometryn Simazin
Pyrethroides	Insecticides	Bifenthrin Cyhalothrin Deltamethrin	Cypermethrin Fenvalerate Fluvalinate Permethrin
Carbamates	Insecticide	Carbaryl	Aldicarb Carbofuran Fenoxycarb Methomyl
	Fungicides	Mancozeb Maneb Thiram	
Organotin	Biocides Antifoulants	Tributyltin Triphenyltin	
Urea	Herbicides	Linuron	Diuron

CATEGORIZATION OF ENDOCRINE DISRUPTORS

Endocrine disruption refers to the ability of chemicals to cause adverse effects and interfere with the body's hormones or chemical messengers, and has raised considerable concern in recent years. The primary function of endocrine system is to maintain a stable environment within the body, referred to as homeostasis. The endocrine system also controls reproduction and growth. A growing body of scientific research indicates that various chemicals could pose a documented risk to wildlife and have the potential to negatively influence human health (Swedenborg et al., 2009). It was recognized, that it is necessary to give a definition of EDCs. The European workshop on the impact of endocrine disrupters on human health and wildlife, held in December 1996 in Weybridge, UK, (European workshop, 1996) proposed the definition of endocrine disruptors, followed by many other alternative definitions. At present, there is a general consensus on the definition of Endocrine Disruptors (WHO document, 2002). It is defined as an exogenous substance or mixture that alters function(s) of the endocrine system and consequently causes adverse health effects in an intact organizm, or its progeny, or (sub)populations. A potential endocrine disruptor is an exogenous substance or mixture that possesses properties that might be expected to lead to endocrine disruption in an intact organizm, or its progeny, or (sub)populations.

A material "Roadmap" entitled "Defining criteria options for identifying Endocrine Disrupters in the context of the implementation of the Plant Protection Product Regulation and Biocidal Products Regulation" was published by the European Commission (EC) (EC document, 2014) in June 2014. Pending the adoption of these criteria, substances that are/or have to be classified, as carcinogenic category 2 and toxic for reproduction category 2 (in accordance with the provisions of Regulation (EC) No 1272/2008) will be considered to have endocrine disrupting properties (Ewence et al., 2015).

The number of substances that have been identified as potential EDCs increases every year and these EDCs belong to different families of chemicals: dioxins, polybromodiphenylethers, phthalates, parabens, alkylphenols, pesticides, etc. In 2007, the EC published a document on the implementation of the "Community Strategy for Endocrine Disrupters – a range of substances suspected of interfering with the hormone systems of humans and wildlife" which was later updated (EC document, 2011). From a total of 564 chemicals that had been suggested by various organizations or in published papers or reports as being suspected EDCs, 147 were considered likely to be either persistent in the environment or produced at high volumes. Of these, however, in the first assessment clear evidence of endocrine disrupting activity was noted for only 66 (assigned Category 1). A further 52 chemicals showed some evidence suggesting potential activity (Category 2). In total 118 substances were categorised in the first exercise of priority setting. Of the 66 chemicals in Category 1, humans were considered likely to be exposed to 60 (EC materials, 2016). Ilustrative list of substances belonging to different types of pesticides classified as Category 1 (endocrinal effect recorded at least on one type of animal); and Category 2 (a record of biological activity *in vitro* leading to disruption) is shown in Table 1.

Substances with endocrine disrupting properties are targeted within the EU Registration, Evaluation, Authorisation and Restriction of Chemicals (REACH) Regulation (REACH 1907, 2006).

According to Endocrine Disruptor Screening Program (EDSP) of US Environmental Protection Agency (US EPA) for approximately 87,000 compounds (Chang et al., 2009), 67 target compounds were confirmed as EDCs, including hormones (natural and synthetic estrogen or steroids), pharmaceuticals and personal care products (PPCPs), industrial chemicals, pesticides, and surfactants. Thereinto, more than two thirds of them belonged to pesticides and were named as EDPs, e.g., organochlorine, organophosphorus, pyrethroids, carbamates, amides, and anilines (Feng et al., 2016). EPA researchers developed the EDSP for the 21st Century Dashboard (EDSP21 Dashboard) to provide access to new chemical data on over 1,800 chemicals of interest. The purpose of the EDSP21 Dashboard is to help the EDSP evaluate chemicals for endocrine-related activity.

Endocrine Disruption. Occurrence and Impacts of EDPs

Hormones are involved in regulation of several different and important processes in organizms. These include regulation of reproduction,development and metabolism. There are several ways that chemical substances may interfere with the endocrine or immune system in humans and wildlife. EDCs interfere with natural hormones by binding to estrogen and androgen receptors and activate various hormone receptors and then mimic the natural hormone's action (agonist action). EDCs may also bind to these receptors without activating them. This antagonist action blocks the receptors and inhibits their action. Finally, EDCs may also interfere with the synthesis, transport, metabolism and elimination of hormones, thereby decreasing the concentration of natural hormones (Mnif et al., 2011). Figure 1 shows the scheme of main endocrine disrupting actions.

There are numerous incidences where humans or wildlife have been exposed to potential EDPs. Pest control agents are known to be a major source of EDPs. Due to their widespread use in agriculture, in and around residence, pesticide exposure is ubiquitous and continuing and poses a potential health risk (Ghisari et al., 2015). There are varied sources of EDPs. Typical human exposure occurs with environmental contamination of the food chain, especially fresh water fish and meat, contact with contaminated household dust, and occupational exposure (Frye et al., 2011).

Direct application, release from some household products (e.g., pest-proof wool carpet, wood), and track-in by people and pets from outside can also contribute to indoor contamination by pesticides. Exposure through ingestion and/or inhalation of indoor dust may be comparable to corresponding food consumption, especially for younger children.

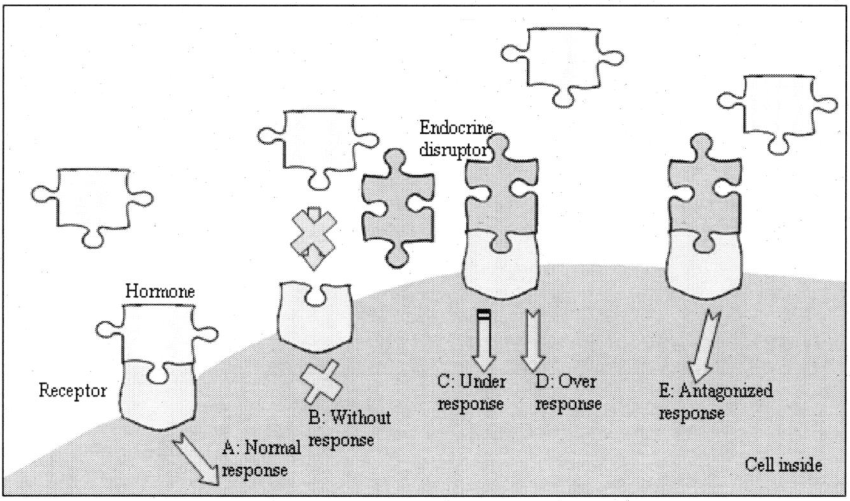

Figure 1. Scheme of the endocrine disrupting action: (A) normal hormonal function, (B) blocking the receptor, (C, D) modify the expressions of specific receptors or (E) antagonize their action.

Pesticides are found at detectable levels in many parts of the environment in the inhabited as well as uninhabited areas of the world. The introduction of EDPs into these areas can occur in several ways and for a variety of purposes.

It can result from direct application to suppress insects and other pests in agriculture, forestry, home, garden, and greenhouse (Rathore et al., 2012). A number of EDPs appear in drinking water as a result of effluent from manufacturing plants entering into streams as well as run-off from agricultural centres (Rogers et al., 2013). Pesticides can enter the soil by direct treatment or by being washed off the plant surface during rainfall (Annamalai et al., 2015). In fact, some EDPs are detectable in so-called "pristine" environments at remote distances from the site where they were produced, used, or released due to water and air currents and via migratory animals that spend part of their life in a contaminated area, to become incorporated into the food chain in an otherwise uncontaminated region (Diamanti-Kandarakis et al., 2009).

The residues of EDPs are found not only in the environment but also in many food items as a result of the ubiquitous nature of pesticide usage with minimal precautions (Kjeldsen et al., 2013). Despite their usefulness in the increase of food production, the extensive use of pesticides during production, processing, storage, transport or marketing of agricultural commodities can lead to the presence of residues in food (Andraščíková et al., 2015). When a crop is treated with a pesticide, a small amount can remain in the crop until after it is harvested (Rathore et al., 2012).

The use of pesticides has led to increased agricultural production but in the meantime induced adverse effects on the human health (Annamalai et al., 2015). EDPs can disrupt the endocrine system and has been associated with advers health effects such as cancer, neurodegenerative disorders and reproductive disorders (Ghisari et al., 2015). Furthermore, they can cause difficulties in cell, organ, individual, and population levels. The negative impact, such as precocious puberty in girls, breast cancer, and reduced sperms are at the forefront of discussions on EDPs. Several chemicals can cause endocrine disrupting effect only in certain species. These materials accelerate damage in specific life periods (embryo, newborn, puberty and pregnancy). Special group – the unborn children, as indirect consumers, is gaining attention as it becomes clear that chemical exposure *in utero* can influence gene expression and thus determine the child's future health and well-being (Shaw, 2009). Fecundity of men was also reported to be decreased in association with occupational exposure to contemporary use of nonpersistent pesticides (Diamant-Kandarakis et al., 2009). In addition, there is evidence that substantiates the effect of EDPs on glucose and lipid metabolism as well as their effect on normal endocrine and hormonal function, via antagonism with hormonal-binding. In addition, binding of organochlorine pesticides (OCPs) to nuclear and cytoplasmic receptors affects the expression of an array of genes encoding for xenobiotic-metabolizing enzymes as well as for cell growth and differentiation (Androutsopoulos et al., 2013). Some EDPs may interfere with the function of hormones, including estrogen, testosterone and thyroid hormones, which are involved in various processes in adults, but also in brain development early in life. These hormones have also been associated with weight homeostasis, both early in development and later in life (Cock et al., 2014). By the damaging of the cholinergic neurons organophosphate pesticides (OPPs) may be responsible for persistent profound neuropsychiatric and neurobehavioral impairments, including memory, cognitive, mental, emotional, motor and sensory deficits (Androutsopoulos et al., 2013).

Representative examples of effects associated with exposure to EDPs in human and wildlife are summarized in Table 2.

It is not uncommon for many classes of pesticides, such as insecticides, herbicides, and fungicides, to be used on the same crop. In the case of insecticides, an adjuvant is often added to the formulations to enhance the intensity of the lethal effect. In the case of herbicides, due to the increasing incidence of plant tolerance to a specific pesticide, some formulations now have as many as three active ingredients. Each active ingredient has a specific mode of action for controlling a pest, and each active ingredient has its own possible side effects on the wildlife and humans exposed to it. It is impossible to determine the cumulative risk posed to wildlife and humans as the result of releasing vast amounts of pesticide mixtures into the environment (Colborn, 2006). Effects of different classes of EDCs may be additive or synergistic because of environmental contamination with mixtures of compounds. The changes caused by the EDCs are thought to be passed down from one generation to the next. For that reason, substances present in the environment in low concentrations seem to pose higher risk to living organizms than short-term high-dose exposure to these chemicals (Diamanti-Kandarakis et al., 2009).

The global increase in concern regarding EDPs results in an urgent need to develop highly specific and sensitive analytical methodologies to test for the presence of EDPs in food, environmental samples, and clinical samples at low concentration levels, lower than required by the legislation. The development of new and improvement of existing methods will offer reliable tools for the control of human and wildlife exposure to EDPs, as a prevention of undesired endocrine disrupting effects.

Table 2. Representative examples of effects associated with exposure to EDPs in human and wildlife

EDPs	Organizm	Effect	Reference
Atrazine	Human hepatoma cells, fish	Increase in protein production; decrease in reproductive ability and immune system response; indirect estrogenic effect via modulation of aromatase expression	An et al., 2014; Kroon et al., 2014
Permethrin	Human hepatoma cells	Increase in protein production; decrease in reproductive ability and immune system response	An et al., 2014
OCPs	Bald eagle plasma, frogs	Insulin resistance; miscarriage; preeclampsia; poor weight gain during fetal development and preterm delivery	Androutsopoulos et al., 2013; Annamalai et al., 2015
OPPs	Human cells	Insulin resistance; reduce the metabolism of oestradiol	Androutsopoulos et al., 2013
p,p-DDE	Human cells, maternal serum	Obesity; increase in leptin release from mature adipocytes and an increase in basal fatty acid uptake	Androutsopoulos et al., 2013; Cock et al., 2014
DDT	*In vitro* studies with preadipocytes, human cells, epidemiological studies	Antiandrogenic effects, reduce the binding of dihydrotestosteron to activated receptor *in vivo*; increase birth weight; incomplete methylation of specific gene regions in the young brain and impaired hippocampal neurogenesis across generations; impaired verbal attention and eye-hand coordination	Androutsopoulos et al., 2013; Cock et al., 2014; Kajta et al., 2013
Methoxychlor	Rat, human cells	Testicular apoptosis; stimulation of ovarian cancer growth	Androutsopoulos et al., 2013; Vaithinathan 2010; Kim et al., 2014
Lindane	Human cells	Androgens antagonist due to inhibition of dihydrotestosteron binding to activated receptor	Androutsopoulos et al., 2013
HCB	Monkey	Increased length of menstrual cycle; decreased progesterone level	
Alachlor	Lizards	Thyroid disrupting effects	Bicho et al., 2013
Vinclozolin	Human breast carcinoma cells	Strong antiandrogenic activity	Christen et al., 2014
Propiconazol	Human breast carcinoma cells	Strong antiandrogenic activity	Christen et al., 2014
Chlorpyrifos	Human cells, rabbit	Stimulate oxidative stress and induce apoptosis and excitotoxicity; alters reproductive function by altering the activity of the pituitarythyroid and pituitary-adrenal axes	Kajta et al., 2013; Mavedati et al., 2015
Fipronil	Chinese hamster ovary cells	Antagonistic activities via estrogen receptor α	Lu et al., 2015
Fipronil sulfone	Chinese hamster ovary cells	Antagonistic activities via estrogen receptor α; anti-thyroid hormone activity	Lu et al., 2015
Endosulfan	Fish	Alteration of the morphometry of gonadotropin -releasing hormones producing cells	Piazza et al., 2015
Monocrotophos	Fish	Hyperplasia and hypertrophy of thyroid follicular epithelium	Zhang et al., 2013

Notes: DDE – dichlorodiphenylethane, DDT – dichlorodiphenyltrichloroethane, HCB – hexachlorbenzene, OCPs – organochlorine pesticides, OPPs – organophosphate pesticides.

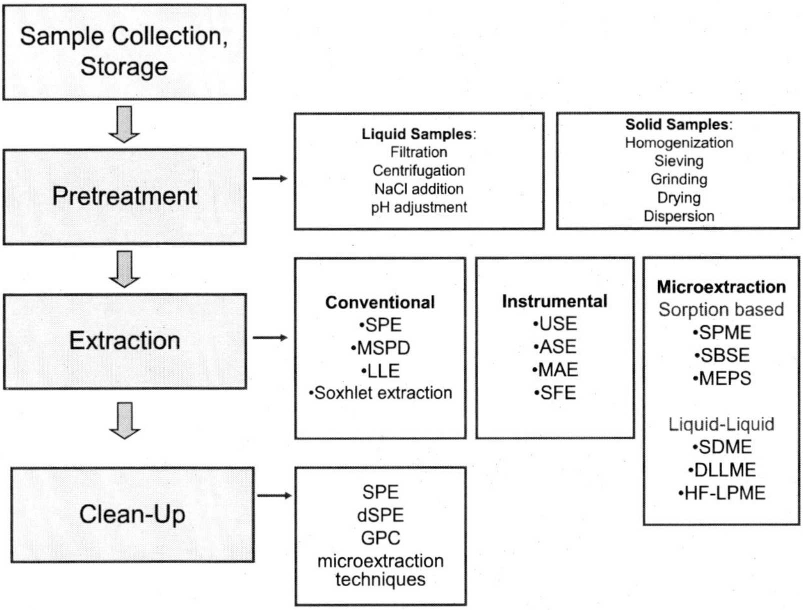

Abbreviations: ASE – Accelerated Solvent Extraction, DLLME – Dispersive Liquid-Liquid Microextraction, GPC – Gel Permeation Chromatography, HF-LPME – Hollow Fibre Liquid Phase Microextraction, LLE – Liquid-Liquid Extraction, MAE – Microwave Assisted Extraction, MEPS – Microextraction by Packed Sorbent, MSPD – Matrix Solid Phase Dispersion, SBSE – Stir Bar Sorptive Extraction, SDME – Single Drop Microextraction, SFE – Supercritical Fluid Extraction, SPE – Solid Phase Extraction, SPME – Solid Phase Microextraction, USE – Ultrasound Assisted Extraction.

Figure 2. Scheme of sample preparation steps within an analytical methodology.

SAMPLE PREPARATION TECHNIQUES

Sample preparation techniques have the same major objectives – isolate the component of interest from the sample; remove potential interferences from the sample matrix; if necessary, convert the analytes into a more suitable form via derivatization or pH adjustment; increase the concentration of the target analytes at concentration detectable by the selected analytical instrumentation; and provide a robust, reproducible method. Sample treatment is considered as the bottleneck of the analytical methodologies in use for the analysis of food samples. It is estimated that 60−80% of the work activity and operating cost in analytical laboratories is spent on preparing samples for introduction into the analytical system. It is also estimated that this part of the analytical process is responsible for more than 50% of the errors associated with the final reported data (Ramos, 2012).

Typical sample preparation steps include sample collection and storage, sample pretreatment, extraction and cleanup. Figure 2 schematically depicts the important steps of sample preparation with most often utilized techniques at each particular step listed.

There is an urgent need to develop analytical approaches that are faster, more automated, cost-effective, and greener alternative to the traditional sample preparation protocols.

This subchapter is devoted to the basic principles of modern extraction techniques, comparing their advantages and drawbacks, and their ability and applicability to EDPs determination, with emphasis on food and environmental samples. The first part describes the conventional techniques, mainly the most widespread techniques – liquid-liquid extraction and solid-phase extraction. Instrumental sample preparation techniques offer the enhanced extraction efficiency, less laboriousness, automatization and higher throughput. Recently emerged QuEChERS (quick, easy, cheap, effective, rugged, and safe) and a variety of microextraction techniques give numerous advantages.

Sample Pretreatment

Samples with complex matrices, containing a lot of sludge, lipids, proteins, salts, sugars, and other unwanted materials, ordinarily require sample pretreatment. Sample pretreatment is sample/analyte dependent. Depending on sample state, it usually comprises filtration, centrifugation, ionic strength and pH adjustment. Solid samples require homogenization, blending, sieving, grinding, and dispersion.

Ultra-trace and trace level quantitation presents challenges with respect to maintenance of analytical equipment in order to obtain reliable results. Larger volumes/quantities of sample are desirable in order to lower detection limits, but the higher the volume/quantity of sample that is pre-concentrated, the more likely the matrix will interfere (LaFleur and Schug, 2011).

Special treatment and care is required for clinical samples, e. g. tissues or blood collected during surgery must be kept frozen, e.g., at −80°C (Arrebola et al, 2012). The quantity of these samples available for testing is usually tiny, therefore, only a limited number of methods is appropriate for this kind of sample preparation.

Special attention must be paid to the laboratory material, its cleaning and storage. According to Bizkarguenaga et al., 2012, glassware is rinsed with water and then soaked in acetone for 24 h. Afterwards the material is rinsed with Milli-Q water. Then all glassware, except for the volumetric one, is dried in the oven at 120°C for at least 4 h, glass tubes are dried in a muffle at 400°C for at least 4 h. Water samples are usually collected in pre-cleaned amber glass bottles, sometimes rinsed with the same water sample and transported to the laboratory in cooled boxes or stored at room temperature until analysis (Bono-Blay et al., 2012). The analyses should be performed within short controlled time period after sampling.

Conventional Sample Preparation

Liquid-liquid extraction (LLE) in conventional shake-flask arrangement, solid-liquid extraction (SLE) and solid-phase extraction (SPE) are the typical representatives of the conventional sample preparation techniques able to achieve separation of EDCs from various matrices. Traditional methods have lots of disadvantages.The procedures are labour-intensive, time consuming, complicated, expensive, require large amounts of organic solvents, which results in generation of hazardous laboratory waste, and usually involve many steps, potentially leading to loss of some analyte.

Liquid-Liquid Extraction

Extraction by solvent is traditionally important technique for the extraction of EDPs from samples. If samples are liquid, technique is called liquid-liquid extraction. For solid matrices, solvent extraction is the most common name. Solvent extraction supported by shaking is based on the partitioning of analytes between liquid and solid phases (in the case of classical LLE between two immiscible liquids), which occurs when analytes move from the sample into the properly selected solvent. This method has many pitfalls, it is laborious and time-consuming, expensive and prone to form emulsion, it requires large quantities of the sample, evaporation of large volumes of solvents and disposal of toxic and flammable chemicals. The most often used solvents today include acetonitrile (MeCN), ethyl acetate and acetone. Toluene is advantageous for proper chromatographic behaviour (Lestremau et al., 2014). The chlorinated organic solvents are effective for the extraction, but from the point of view of health and environmental safety, the negative impact is very high. Chlorinated solvents are still used, but in combination with other solvents, such as *n*-hexane: dichloromethane (Sichilongo and Banda, 2013) or dichloromethane: acetone (Uygun and Dilgin, 2013).

There is limited potential to adjust the selectivity of the extraction; therefore, extensive clean-up is required prior to subsequent chromatographic separation. The most often used clean-up is realized by SPE (Li et al., 2013, Pelit and Dizdas, 2013) or by gel permeation chromatography (GPC) (Fujii et al., 2011).

Solid-Phase Extraction

SPE is widely used technique and was primarily developed as the alternative to LLE. Within the advantages of SPE over LLE it is necessary to mention the higher degree of automation, higher throughput, and lower solvent consumption.

For SPE, a sorbent will retain and concentrate some target analytes from the sample solution due to strong affinity between sorbent and analytes. The other purposes of SPE could be the clean-up for purification of the extract, and in some cases for solvent exchange and derivatization of analytes.

SPE is commonly arranged as the extraction by sorbent placed in disposable cartridges or disks for large volume sample preparation. After first conditioning the sorbent, the liquid sample is loaded onto the cartridge. The analytes are retained on the sorbent bed; interferences are eluted or washed in the washing step. At the end, analytes are eluted using a small volume (few mLs) of a suitable solvent. If the SPE is used to remove interferences from the matrix, sorbent or mix of sorbents is added to the crude extract. This sub-technique is called dispersive solid-phase extraction (dSPE).

Selection of the sorbent is the most important parameter in method development. SPE was mainly applied to extract low-levels of EDPs employing cartridges containing reversed-phase or normal phase SPE sorbent such as octadecyl (C18) or octyl (C8) silica bonded phases (e.g., Supelclean ENVI-18, Lichrolut EN RP-18), ion-exchange sorbents and mixed-mode sorbents, polymeric sorbents (Strata X, hydrophilic-lipophilic balance polymer (HLB), styrene divinylbenzene copolymer Bond Elut-ENV) and their combinations. Carbon sorbents (Envi-Carb) are occasionally used for environmental sample preparation. Polymer-based SPE involves macroporous polymeric media, offering a higher loading capacity than conventional functionalized silica SPE media. Oasis HLB cartridge, with a hydrophilic–lipophilic balanced copolymer of n-vinylpyrrolidone and divinylbenzenes, with good retention for both polar and nonpolar compounds, was used for multiresidue analysis of multiclass contaminants in water samples (Bizkarguenaga et al., 2012; Bono-Blay et al., 2012; Oguz and Kankaya, 2013) or rat serum (Tournier et al., 2015). For more details, see Table 3.

For clean-up purposes, the most popular sorbent is florisil, combination of primary secondary amine (PSA) and carbon sorbents, namely graphitized carbon black (GCB). The selection of proper sorbent material depends on a series of factors that include the kind of sample, sorbent selectivity required, cost, etc.

Recently, much progress has been made towards development of new advanced materials for sorption purposes and improvements of properties of commonly used sorbents, such as molecularly imprinted polymers (MIPs), restricted-access materials (RAMs), immunoaffinity sorbents (IASs), carbon nano-tubes and different tailor-made sorbents (Buszewski and Szultka, 2012).

Matrix Solid Phase Dispersion

Matrix solid phase dispersion (MSPD) is an extraction and clean-up procedure used for the simultaneous dispersion and extraction of viscous liquids, semisolid and solid samples. MSPD was developed by Barker, 2000 and recent developments were reverviewed and updated by Capriotti et al., 2010. MSPD is performed by blending the sample with an appropriate amount of solid support materials (e.g., Florisil or silica gel modified by C8, C18) in a glass mortar with a glass pestle, to generate a chromatographic material suitable for the extraction of compounds from the dispersed sample. Florisil is used to isolate more polar analytes, whereas non-polar modified silica (C18) is usefull for non-polar analytes. Subsequently, the homogenized mixture is packed into a syringe barrel SPE column, from which target compounds are isolated with a suitable elution solvent. MSPD offers the possibility of performing extraction and clean-up in one step and does not need special instruments.

The advantage of MSPD is its simplicity, speed, and the elimination of emulsions formation. The MSPD was used to extract 14 OCPs and 7 polychlorinated biphenyls (PCBs) in edible vegetable oils, where 0.5 g of oil sample was blended with 3.5 g of sulfuric acid-impregnated silica and 0.8 g of silica gel was used as co-column sorbent. (Li et al., 2014).

Instrumental Sample Preparation

Traditionally, Soxhlet extraction was used for the extraction of pesticides in solid matrices, whereby the sample is placed in a porous thimble and is continuously extracted in a glass apparatus with a sub-boiling solvent. The timble serves also as a filtration medium. Soxhlet methods are simple, standard; however, they are slow, inefficient and they consume high volumes of organic solvent. Pesticide residues extraction from solid samples is a difficult task. To overcome the drawbacks of traditional approaches, many assisted methods with enhanced efficiency have been developed, such as ultrasound assisted solvent extraction (USE), accelerated solvent extraction (ASE), supercritical fluid extraction (SFE), and microwave assisted extraction (MAE). The enhanced extraction efficiency is due to ultrasound, higher pressure, and microwave energy, which increase analyte release from the solid matrix. Recently published book covered the topic of enhancing extraction processes in food industry (Lebovka et al., 2012).

Ultrasound Solvent Extraction

For USE, acoustic energy is applied to enhance the efficiency of solvent extraction. Acoustic energy forces the elimination of analytes from the structure of the matrix. USE provides a more efficient contact between the solid sample and solvent due to increase of both pressure (which favours penetration and transport) and temperature (improves solubility and diffusivity). The latest advances in ultrasonic devices along with their most recent applications for analytical chemistry have been reviewed by Santos and Capelo, 2007.

The typical extractants are methanol, ethanol, MeCN, and acetone in the mL range of volume, and the sonication times lie in the range of 2–120 min. After sonication, the extracted analytes are separated from the matrix by vacuum filtration or centrifugation. The process is repeated two or three times to achieve higher extraction efficiency; the extracts are then recombined, and a clean-up procedure with SPE cartridges is applied prior to the analysis (Sosa Ferrera et al., 2013).

Several extractions can be performed simultaneously and ultrasound laboratory equipment is relatively inexpensive and commonly present throughout analytical laboratories.

The selectivity of the extraction is low and after the first extraction, purification and pre-concentration step is commonly applied. Novel dispersive liquid–liquid microextraction was shown to be effective for pre-concentration of OPP residues analysis after USE from tomato (Bidari et al., 2011). USE combined with hollow-fiber liquid-phase microextraction prior to GC analysis was presented by González-Curbelo, 2013 for determination of OPPs and metabolites in cereal-based baby foods and wheat flour.

Accelerated Solvent Extraction

ASE, also called Pressurized Liquid Extraction (PLE), uses conventional solvents at elevated temperatures and pressures, which results in improved extraction kinetics to achieve quantitative extraction from solid or semisolid samples in short time with a small volume of solvent. Pressure keeps the solvent below its boiling point, and in solid samples forces its penetration into the pores of the sample and thus enhances the extraction (Beyer and Biziuk, 2008) and high recoveries. The parameters affecting ASE efficiency and application advancement of ASE in the analysis of organic contaminants, natural toxic compounds as well as bioactive and nutritional compounds in animal origin food, plant origin food and animal feed were reviewed in detail by Sun et al., 2012. Very recent review by Vazquez-Roig and Pico, 2015 evaluated the application of PLE to the extraction of organic contaminants in environmental and food samples.

The advantages of ASE over traditional methods like Soxhlet extraction is the possibility to reduce the extraction time and the solvent consumption, high level of automation and thus higher throughput. The upper extraction temperature under pressure conditions is not limited by the boiling point of the solvent as it is in traditional Soxhlet extraction. The undesired dilutions of the analytes and the high initial necessary investment to the instrument are the main limitations of ASE. The organic solvent consumption can be minimized if pressurized hot water extraction (PHWE) is chosen, using pure water with the addition of a water miscible organic solvent, if required. Selective pressurized liquid extraction (SPLE) is an analytical technique that builds upon PLE by incorporating matrix-compound (i.e., interference) retainers into the extraction step, thereby reducing sample-preparation steps and increasing sample throughput (Subedi et al., 2015).

The most common organic solvents for ASE of EDPs in food and environmental samples are acetone, MeCN, ethyl acetate, hexane and combinations of above mentioned solvents with chlorinated solvent. However, for most hydrophobic compounds, the use of non-polar organic solvents is unavoidable to achieve an efficient extraction, and an additional extract clean-up step is often mandatory to remove the interferences (co-extracts such as sulfur and lipids) prior to analysis of the samples. Evaporation of extracts and use of SPE or GPC is a common approach for post-extraction treatment. An alternative option to minimize time and solvent consumption is performing both extraction and purification stages simultaneously by placing adequate sorbents inside PLE cells (in-cell clean-up). This reduces the need for post-clean up procedures and allows the automation of clean-up steps, a key factor in avoiding excessive sample manipulation. Simultaneous determination of 97 contaminants, including fragrances, UV filters, repellents, endocrine disruptors, biocides, polycyclic aromatic hydrocarbons (PAHs), PCBs, organophosphorus flame retardants, and several types of pesticides in marine sediment samples using alumina as an in-cell clean-up sorbent serve as an example of this approach (Pintado-Herrera et al., 2016).

Microwave Assisted Extraction

The procedure of MAE involves placing the sample with the solvent into special tube and heating the solvent by microwave energy. This process ensures favourable partitioning of analytes from the matrix to the solvent and consequently, leads to the reduction of analysis time. MAE systems can operate in two modes: open (focused MAE) or closed (pressurized MAE) vessels. In the latter devices, the solvent is heated and pressurized.

Main advantages of MAE are reduction of solvent volumes, enhanced extraction efficiency, shortening of extraction time, and increased throughput of samples. Due to the lack of extraction selectivity, the clean-up of crude extract is usually applied, e. g. by dSPE (Satpathy et al., 2011). MAE is mostly applied to the analysis of soils and sediments. In recent years, several reviews (Sanchez-Prado et al., 2015; Wang et al., 2016) were published on the use of microwaves for solid-sample extraction.

Supercritical Fluid Extraction

SFE is the process of separating analytes from the matrix using supercritical fluids as the extracting solvent (Lang and Wai, 2001). Principles, potential and pitfalls of the methods in comparison to ASE and MAE were first reviewed by Camel, 2001. Developments of SFE in food science, natural products, by-product recovery, pharmaceutical and environmental sciences were updated in a review by Herrero et al., 2010. SFE can be mainly applied to the extraction of non-polar and slightly polar analytes, primarily applicable to solid samples. Among all the solvents used in SFE, pure carbon dioxide (critical point: 30.9°C and 73.8 bar) is commonly used as fluid because of its low critical constants, its low toxicity and cost, chemical inertness, and its ability to extract quantitatively a wide range of relatively non-polar organic compounds, including those with high molecular masses, from a variety of matrices. CO_2 is environmentally friendly and safe. The lack of extraction efficiency for polar compounds can be overcome through the addition of polar modifiers to CO_2, such as methanol and water.

SFE has gained increased attention as an extraction method because it provides cleaner extracts that can generally be analyzed by GC without any further clean-up step. Valverde et al., 2009 compared the recoveries of 22 pesticides by SFE with those obtained with an ethyl acetate solvent extraction/ GPC clean-up method and obtained consistent results. Recent applications of SFE in the food, beverage and nutraceutical industries were reviewed by Knez et al., 2013.

QuEChERS

The sample preparation method QuEChERS with the name composed as the acronym of words "quick, easy, cheap, effective, rugged and safe" was introduced by Anastassiades et al., 2003. It represents a breakthrough in the field of sample preparation. From the beginning it has been a widely used method for food sample preparation, which was later broadened to various analytes and matrices, including soil or clinical samples.

QuEChERS is based on MeCN extraction followed by dSPE, all simply performed in two centrifugation tubes. 10-15 g of sample is homogenized, and pesticides are extracted by organic solvent, e. g. MeCN, followed by salt-out partitioning of the water from the sample using anhydrous $MgSO_4$, NaCl, and/or buffering agents. The clean-up is performed by dSPE or disposable pipette extraction (DPX) with anhydrous $MgSO_4$ and by variety of sorbents. PSA retains polar interferences, including fatty acids, other organic acids, sugars and pigments, such as anthocyanidines. GCB efficiently removes pigments (chlorophyl and carotenoids), but showed strong affinity towards planar molecules, including pesticides. C18 modified silica is used for the elimination of fatty interferences. Zr-Sep is a mixture of C18 modified silica and silica coated with zirconium-dioxide, with ZrO_2/C18 (in the ratio of 2/5). Zr-Sep$^+$ is silica coated with zirconium-dioxide and octadecyl-siloxan carrier groups (Shimelis et al., 2007. Zr-Sep and Zr-Sep$^+$ are promising materials for clean-up of high lipid content fruit and vegetable extracts, successfully applied individually to avocado and almond samples (Lozano et al., 2014), and to soya-based nutraceuticals (Páleníková et al., 2015[a]). Multi-walled carbon nanotubes mixed with other materials like PSA, GCB and C18 modified silica, showed brilliant cleanup performance in multi residue pesticide monitoring (Qin et al., 2015).

The QuEChERS method is relatively new method, but it quickly gained general acceptance by scientific standards organizations – acetate version AOAC Official Method 2007.01 and citrate version European Committee for Standardization (CEN) Standard Method EN 15662. Comparison of un-buffered and both buffered methods (Lehotay et al., 2010) was evaluated in terms of pH-dependence, co-extracted matrix components, matrix effects and substitution of MeCN with ethyl acetate. Differently buffered methods were compared for high acid content matrices (Andraščíková and Hrouzková, 2013).

Recovery tests are ordinarily assayed at two fortification levels, 10 and 100 μg/kg. In the case of endocrine disruptors it is good to enhance the selectivity of the system in order to investigate at lower levels, such as 1 and 50 μg/kg in three matrices, tomato, apple, and orange (Belmonte Valles et al., 2012), or 0.1, 1, 5 and 250 μg/kg (Hrouzková et al., 2012).

The contribution of our research group to EDP method development in non-fatty foods was focused on the development of the fast GC-MS method for separation, detection and quantification of EDPs belonging to different chemical classes – OCPs, OPPs, pyrethroids, dicarboximides, phtalamides, dinitroanilines, pyrazoles and triazinones. The developed method involves the QuEChERS sample preparation, modified according to needs and resources (Hercegová et al., 2010, Húšková et al., 2010[a]). Using this method, a survey of EDPs in different fruit and vegetable matrices was performed (Hrouzková et al., 2012).

Several papers reviewing QuEChERS have been published. Wilkowska and Biziuk, 2011 reviewed the applications to food matrices, Hercegová et al., 2007[a,b] focused on baby food. The newest one published by González-Curbelo et al., 2015 was devoted to the evolution of QuEChERS and its application fields, including not only pesticide analysis, but also pharmaceutical, mycotoxin and PAHs analysis.

Microextraction Techniques

Solid-phase microextraction (SPME) and stir bar sorptive extraction (SBSE) are the most often employed representatives of sorption–based microextraction techniques. Single drop microextraction (SDME), dispersive liquid-liquid microextraction (DLLME), hollow fiber liquid phase microextraction (HF-LPME) and many modifications and variations are typical techniques based on microextractions by liquid phase. Owing to numerous advantages, microextraction methods are attracting the attention of researchers.

Sorption-Based Microextractions

SPME is a miniaturized technique. In SPME, the analytes are sorbed onto a fused-silica fiber coated with an appropriate sorbent layer by simple exposure of the fiber for a preselected time to the headspace (HS) of the sample or by direct immersion in a liquid sample. Since its introduction by Pawliszyn's group (Arthur and Pawliszyn, 1990) as a solvent free preconcentration technique, SPME has been widely used in many application areas. SPME is an alternative extraction method to traditional techniques, due to no solvent consumption in combination with GC, blanks reduction and a decrease in extraction time. This method does not require complete removal of the analyte from the liquid matrix and can be used for a wider range of applications in comparison with other techniques such as SPE, which requires an exhaustive extraction (Katoka et al., 2009). The major advantages of SPME are its easy miniaturization, automation of devices, and its convenience in coupling with chromatographic instruments. Several coating fibers are commercially available for SPME, including polydimethylsiloxane (PDMS), polyacrylate (PA), divinylbenzene (DVB), carboxen (CAR), carbowax (CW), and their copolymers (Luo et al., 2012). The developments of new extraction phases greatly benefit from sol-gel technology (Souza-Silva et al., 2015), which is undoubtedly one of the most ingenious ways of creating tailor-made extraction phases. Ke et al., 2013 immobilized the graphene onto a stainless steel wire to obtain a novel SPME fiber coating based on sol–gel technique for OCPs extraction from water. The new fiber possessed a homogeneous, porous surface and showed excellent thermal and solvent stability. A robust SPME fiber was fabricated by electropolymerization of thiophene on a stainless steel wire (Pelit and Dizdas, 2013). Polythiophene coatings display higher extraction efficiency toward some analytes in comparison to other conducting polymers. The new fiber was applied to the determination of volatile EDPs, namely, chlorpyrifos, penconazole, procymidone, bromopropylate, and λ-cyhalothrin in well waters by a HS-SPME.

SBSE, first proposed by the Sandra's research group in 1999 (Baltussen et al., 1999), is based on sorptive extraction, whereby the solutes are extracted into a polymer coating on a magnetic stirring rod. SBSE works on the same general principles as SPME. Both techniques use in many cases the same extraction phase a PDMS layer, but the amount of PDMS of SBSE is 50–250 times greater than that used in SPME. Initially SBSE was mostly used for the extraction of compounds from aqueous matrices. The technique has also been applied in HS mode for liquid and solid samples and in passive air sampling mode. The stir bar could be thermally desorbed on-line in heated injector of gas chromatograph rendering this technique completely solventless. Recently, Prieto et al., 2010 reviewed SBSE optimisation, limitations, potential solutions such as in-house coatings and derivatisation and novel applications in multi-residue analysis and passive sampling. Kawaguchi et al., 2013 and David, 2015 overviewed applications of SBSE to the food analysis and Camino-Sánchez et al., 2014 discussed applications, limitations and future trends in SBSE. Main limitations of SBSE technique are long extraction and desorption times. An extraction time of 14 h was reported by Chary et al., 2012 for endocrine-disrupting chlorinated compounds in river water and wastewater effluent.

Microextraction by packed syringe (MEPS) is a newly introduced technique based on the principle of miniaturized SPE (Abdel-Rehim et al., 2004, Abdel-Rehim, 2011). MEPS works with small-volume samples (approximately 10–250 μL). A small amount of a sorbent (approximately 1–4 mg) is packed into the microsyringe barrel or a container between the needle and the barrel, which is used as a mini SPE cartridge. Common arrangement employs 100 μL gas-tight syringe equipped with a small container incorporated into the needle. The applications in various areas were recently reviewed by Moein et al., 2015 and Páleníková and Hrouzková, 2014. Two one-step extractions were chosen and compared: the proven conventional SPE technique for the extraction of biological fluids and a new technique MEPS. Even though MEPS technique is solvent-saving (a ‘‘green” method as widely named), it was not appropriate for multifamily analytes including conjugated hormones. Instead, polymeric-based SPE was used for the mix of endocrine disruptors and endogenic androgens and estrogens in rat serum (Tournier et al., 2015).

Liquid-Liquid Microextraction (LLME)

Liquid-liquid microextraction (LLME) is a set of novel solvent-minimized extraction techniques where only microliters of solvent, as opposed to hundreds of millilitres needed in conventional LLE, are used to concentrate analytes from various samples. Asensio-Ramos et al., 2011, Sarafraz-Yazdi and Amiri, 2010, Andraščíková at al., 2015 recently published a detailed review of principles and applications of LLME techniques to the extraction of analytes from various matrices. Very small volumes of extraction solvent can be either directly immersed in (direct extraction) or suspended above the sample (HS extraction), and directly introduced for subsequent instrumental analysis.

SDME is based on the distribution of analytes between a microdrop of extraction solvent (usually few microliters) at the tip of a microsyringe needle and a liquid sample containing the analytes. After extraction, the microdrop is retracted back into the microsyringe and injected into chromatographic or electrophoretic system for further analysis (Zhao et al., 2006). In this way, high enrichment factors are obtained owing to the high ratio of sample volume to organic phase volume and, consequently, low limits of detection (LODs) can be obtained. SDME has many advantages such as low cost, simplicity, speed, low environmental impact and compatibility with GC and high performance liquid chromatography (HPLC) injection systems as well as its applicability to different matrices, since there are a good number of solvents that can be used. In this sense, toluene, hexane, cyclohexane and xylene are among the most common, although ionic liquids also provide good results due to more reproducible extraction conditions as well as larger and more stable drops allowing prolonged extraction times (Asensio-Ramos et al., 2011). Reproducibility of the extraction is often poor due to significant dissolution of organic extractant, which has a small volume and large contact area with sample solution. The other important limitation of SDME is the instability of the drop as it is easily dislodged from the tip of the microsyringe needle due to prolonged extraction time, fast stirring rate and more complex matrix. In order to overcome the drawbacks of SDME, modifications of the syringe needle tip were attempted (Pakade and Tewary, 2010). The addition of a silicone ring in the microsyringe needle improved the stability in a larger microdrop volume (Fernandes et al., 2012). Solidification of floating organic drop microextraction (SFODME) has been introduced by Khalili Zanjani et al., 2007 as a modification of SDME, based on liquid phase microextraction principle. In this method, a micro drop of an appropriate solvent, e. g. 1-undecanol, which has the melting point near room temperature, is used as an extraction solvent. The solvent drop is transferred to the surface of the water sample, which is being agitated by a stir bar. After the extraction, the sample vial is cooled down and the solidified micro drop is transferred into a conical vial, where it melts immediately and it is injected to gas chromatograph (Pelit and Yengin, 2014).

In HF-LPME technique a hollow fiber unit acts as the holder of microliter amounts of organic solvent. Two modes of HP-LPME technique, two-phase and three-phase were developed. Pedersen-Bjergaard and Rasmussen, 1999 introduced the use of hollow fibers in a three-phase LPME procedure. The technique utilizes a porous hollow fiber, containing suitable organic solvents inside the pores of the fiber as a membrane between two aqueous phases, the donor (outside) and the acceptor (lumen of fiber). The acceptor solution can be directly injected into the chromatographic instrument. The main advantages of HF-LPME are the small volume of organic solvent, low cost, simplicity and a wide range of applicability to pesticide residues in food or environmental samples. Some disadvantages of LPME using hollow fiber membranes, are as follows: (1) existence of a membrane barrier between the source (sample) phase and receiving (acceptor) phase reduces extraction rate and increases extraction time; (2) in two phases LPME excess amount of solvent is needed for elution of analytes from lumen and pores of fiber. This process is also a time consuming step; (3) creation of air bubbles on the surface of the hollow fiber reduces the transport rate and decreases the reproducibility of the extraction; (4) in real samples such as blood plasma, urine, wastewater, etc. adsorption of hydrophobic substances on the fiber surface may block the pores (Khalili Zanjani et al., 2007).

DLLME is based on the ternary component solvent system (aqueous sample, dispersive solvent and extractive solvent). A mixture of an extractive solvent (organic solvent or ionic liquid) and a dispersive solvent (water-organic miscible solvent) is rapidly injected into an aqueous sample, thereby forming a cloudy solution with plenty of tiny droplets for efficient extraction. The extractant is then settled and centrifuged; the analytes are isolated and preconcentrated in the organic extractive phase (Rezaee et al., 2010). Among the advantages of DLLME are usage of small volume of organic solvents, easy operation, speed, low cost, high recovery, high enrichment factor and easy linkage to analytical methods. DLLME is easily applied to analysis of liquid samples, such as fruit juices, as reviewed recently by Ahmad et al., 2015 and Andraščíková et al., 2015.

Separation Techniques

Identification of endocrine disruptors requires analytical methods that are capable of detecting large numbers of compounds in a single analysis. Many studies reported various analytical methods for the determination of EDPs (see Table 3). Most of the EDC detection and quantification is based on GC and/or HPLC methods thanks to their high separation efficiency and selectivity that give the primary information about the physico-chemical properties to classify the sample composition (Locatelli et al., 2016).

Gas Chromatography

GC is the method of choice when dealing with completely unknown compounds. GC-amenable compounds should vaporize at temperatures below 350°C. The technique is applicable to compounds ranging from permanent gases up through hydrocarbons as large as 100 carbons with molecular masses less than 500 atomic mass units (amu). It is used for lipophilic compounds with $pK_{ow} > 2$.

GC is considered a powerful technique for the separation, identification and quantification of organic compounds at low-concentration levels (Mansilha et al., 2010), but it also has some disadvantages (Bizkarguenaga et al., 2012). Analytes that are too polar, non volatile or thermolabile (e.g., carbamates), often undergo the breakdown in the injection port or in the column during the analysis. Therefore, derivatization by various derivatization reagents is desirable prior to GC analysis to increase the chromatographic response (Yang and Shin, 2013).

Split/splitless inlet is the classical solution for sample injection. Injector in the splitless mode is the most commonly applied in EDP residues analysis (see Table 3). However, splitless injection has some limitations in terms of small sample capacity, discrimination of analytes, and adsorption and degradation of pesticides, which can negatively influence the quantitative analysis. Therefore, programmed-temperature vaporization (PTV) injector is at present the recommended choice for sample introduction protecting against effects of co-extracted compounds. Operating in solvent vent mode, PTV allows larger sample volume introduction than in splitless mode, resulting in excellent limits of quantification (LOQs) (Húšková et al., 2010[a], Hrouzková et al., 2012). PTV significantly eliminates matrix effects by releasing high-boiling co-extracted compounds through the split vent and/or trapping in a liner. The application of large volume injection (LVI) in a PTV results in decreases detection limits (Fernandes et al., 2012; Bizkarguenaga et al., 2012). On the other hand, clean-up of complex extracts becomes more important in LVI in order to avoid adsorption and contamination of the injection liner, capillary column, and MS source by non-volatile matrix constituents (Matisová and Hrouzková, 2012).

For separation of multicomponent mixtures, conventional capillary GC with internal diameter (I.D.) of 0.20 – 0.32 mm provides analysis time in the range of tens of minutes. The column parameters such as length, internal diameter or film thickness are parameters to optimize depending on the number of analytes to be separated and the required separation efficiency. Non-polar columns with 5% phenyl 95% polydimethylsiloxane stationary phase are the most often applied in multi-residual analysis. However, medium-polarity (50% phenyl 50% dimethylpolysiloxane) and high-polarity columns are applied depending on the polarity of analytes. Helium is the most frequently used carrier gas; although there have been attempts to replace helium with hydrogen, due to the worldwide lack of helium.

Fast GC yields faster analysis times than conventional GC, resulting in decreased costs, higher labaoratory throughput as more samples could be analyzed, while the separation efficiency remains the same, and, at present, it does not require any additional instrumental equipment. The most common approaches to fasten the separation are the following: *(i)* the use of narrow-bore columns, *(ii)* fast temperature programming, *(iii)* low-pressure gas chromatography (LP-GC), *(IV)* comprehensive GCxGC (Hrouzková and Matisová, 2011).

Concerning detection, the most widely used detection technique of EDPs has been MS. However, the electron capture detector (ECD) has also been frequently used for the detection of pesticides (Li et al., 2014; Pelit et al., 2013; Korba et al., 2013; Arrebola et al., 2012). Unfortunately, specific detectors such as ECD do not provide enough information to confirm the presence of the analyte. MS detector is needed for unambiguous identification of pesticide residues, as defined by SANTE document issued by EC (SANTE document, 2015).

Liquid Chromatography

Until recently, GC-MS was the most widely used technique to analyze residues in complex matrices. HPLC was applied most successfully to the analysis of thermally unstable pesticides. However, the pesticides used today are more polar, thermally unstable or not easily vaporized, which makes them more amenable to analysis by LC-MS.

HPLC has become a popular tool in trace analysis of polar and degradable toxic compounds and also their even more polar degradation products mainly in the pharmaceutical and forensic sciences. This technique is the method of choice for quantifying polar hydrophilic compounds ($pK_{ow} < 2$) and compounds with higher molecular masses including pesticides.

Most studies dealing with the separations of multicomponent pesticide mixtures by HPLC were carried out on columns with dimensions: length of 50 – 250 mm, internal diameter in the range of 2 – 4.6 mm I.D., and 5 μm particle size. Smaller particle sizes such as 1.5 – 2 μm are used in HPLC termed high speed or ultra(high)performance LC (UHPLC). The main advantage of UHPLC is the possibility to achieve ultra-fast and/or high-resolution separations, with reduced solvent consumption; using chromatographic systems compatible with pressures up to 1400 bar (Fekete et al., 2014).

A gradient programme is ordinarily used with the mobile phase, combining various solvents, including water, MeCN, or methanol, to obtain high selectivity and sensitivity. The selection of the mobile phase depends on many factors. Mobile phase plays a critical role in the ionization efficiency; therefore, the influence of different buffers such as ammonium formate mixed with formic acid, ammonium acetate mixed with formic acid with different concentrations of formic acid should be taken into a consideration (Li et al., 2013). The common stationary phases are octadecyl silica (Laborie at al., 2016; Tournier et al, 2015), and octyl silica (Maggioni et al., 2013). The use of a monolith-sorbent-based instead of porous particle-based column packings has become popular over the past few years. Monoliths can accept high flow rates (up to 10 mL/min) in conventional column lengths without generating high backpressures, which is their main advantage. Monolithic rods are prepared by sol-gel technology, which enables the formation of a highly porous material that contains both macropores and mesopores in its structure. The resolution of the monolith rod column is significantly less affected by particulate materials (Sosa-Ferrera et al., 2013). Hydrophilic interaction LC (HILIC), recognized as a special trend in the advancement of HPLC, is used for separating polar compounds that elute in order of increasing polarity (Li et al., 2012).

Only a few studies have used methods employing HPLC detectors other than MS for the analysis of EDPs. The use of fluorescence detector (FLD) or diode array detector (DAD) has been reported in the literature. The whole subchapter will be devoted to mass spectrometric detection.

Mass Spectrometric Detection of EDPs

Detection of EDPs and subsequent screening requires sensitive and selective instrumental analytical techniques with sufficiently low limits of detection and determination. The most efficient approach to EDPs analysis involves the use of chromatographic methods for separation prior to the MS detection. Selective detectors utilized in the past like ECD, nitrogen-phosphorus, flame-photometric, pulsed flame photometric in GC or ultra violet (UV), DAD in HPLC do not provide unambiguous identification. Mass spectrometry is considered as one of the most valuable detection techniques widely used in food safety and quality because it provides information about the molecular structure of the compounds; it is highly sensitive, selective, with high throughput (SANTE document, 2015).

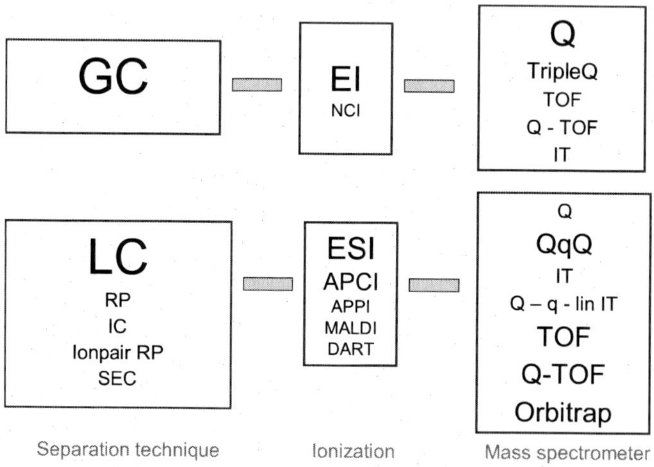

Abbreviations: APCI – atmospheric pressure chemical ionization, APPI – atmospheric pressure photoionization, DART – direct analysis in real time, EI – electron ionization, ESI – electrospray ionization, IC – ion chromatography, IT – ion trap, lin IT – linear ion trap, MALDI – matrix assisted laser desorption ionization, NCI – negative chemical ionization, Q – quadrupole, q – quadrupole as a collision cell, RP-LC – reversed phase liquid chromatography, SEC – size-exclusion chromatography, TOF – time of flight.

Figure 3. Typical combinations of chromatographic separation technique, ionization mode and mass spectrometric detection techniques.

The most common combinations of chromatographic technique with various ionization modes and mass spectrometric analyzers for EDPs analysis are outlined in Figure 3.

Table 3. An overview of analytical methods for analysis of EDPs

Analytes	Matrix	Extraction/ clean-up	Injection technique	LOD	Separation and detection technique	References
Food						
20 OCPs	9 vegetable matrices	SBSE (PDMS)	LVI - PTV, SVV	< 10 μg/kg	GC-MS Q (SIM) EI	Barrida-Pereira et al., 2010
29 pesticides	Fruit and vegetables	QuEChERS/ dSPE	PTV, SVV	≤ 5 μg/kg	fast GC-MS Q (SIM) EI	Húšková et al., 2010[a] Hercegová et al., 2010
35 pesticides	Fruit and vegetables	QuEChERS/ dSPE	PTV, SVV	EI: ≤ 5 μg/kg, NCI: ≤ 1 μg/kg	fast GC-MS Q (SIM) EI, NCI	Hrouzková et al., 2012
53 pesticides	Fruit and vegetables	QuEChERS (EtAc)	splitless	< 0.1 μg/kg	GC-MS/MS QqQ NCI	Belmonte Valles et al., 2012
19 pesticides	Orange	QuEChERS/ DLLME	splitless	0.02-47 ng/g	GC-MS Q (SIM) EI	Andraščíková et al., 2013
14 OCPs	Strawberry, strawberry jam and soil	SDME (n-hexane)	PTV LVI	0.02-0.15 μg/kg	GC-MS/MS QqQ (SRM) EI	Fernandes et al., 2012
19 triazine pesticides	Cereal samples	LLE (MeCN saturated with petroleum ether)/SPE (Oasis MCX)	valve	0.0024-0.42 μg/kg	HPLC-MS/MS LIT (SRM) ESI	Li et al., 2013
40 pesticides	Fish fillet	QuEChERS/ dSPE ($MgSO_4$, C18, PSA)	PTV	0.3-7.5 μg/kg	GC-MS/MS QqQ (SRM) EI	Munaretto et al., 2013

Table 3. (Continued)

Analytes	Matrix	Extraction/ clean-up	Injection technique	LOD	Separation and detection technique	References
15 OCPs	Edible marine biota	SPE (C18)/SPE (NH2)	on-column	0.5-10 ng/g LOQ	GC-MS Q (SIM) EI	Santhi et al., 2012
20 OCPs	Milk	SPE (C18, MNP)	autosampler, not reported mode	0.1-0.3 μg/L	GC-MS Q (SIM) EI	Synaridou et al., 2014
14 OCPs	Edible vegetable oil	MSPD (silica gel)	splitless	0.04-0.74 ng/g	GC-ECD	Li et al., 2014
6 pesticides	Wine	LLE (cyclohexane:EtAc 1:1)/SPE (florisil)	splitless	0.6-2.9 ng/mL	GC-ECD	Pelit et al., 2013
5 pesticides	Wine	SPME (sodium dodecyl doped polypyrrole)	splitless	0.073-1.659 ng/mL	GC-ECD	Korba et al., 2013
4 pesticides and metabolites	Drinking water and PET bottled mineral water	SPE	valve	0.05-7.7 ng/L	LC-MS/MS QqQ (MRM) ESI	Maggioni et al., 2013
21 pesticides	Drinking water	SPE (ENVI-18)	not reported	< 10 ng/L	GC-MS Q (SIM)	Feng et al., 2016
3 pesticides	Drinking water	SPE	valve	0.00005-0.1 μg/L	LC-MS/MS QTrap (MRM)	Odendaal et al., 2015
Environmental samples						
32 EDCs and pesticides	Water	SPE (LiChrolut EN/RP-18, Strata X)	splitless	5.3-95.9 ng/L	GC-MS/MS QqQ, (MRM), EI	Mansilha et al., 2010

Analytes	Matrix	Extraction/ clean-up	Injection technique	LOD	Separation and detection technique	References
76 micropollutants (3 pesticides)	Water	SPME (PDMS/DVB)	spitless	0.0005-0.1 µg/L	GC-MS Q (FS) EI	Martinez et al., 2013
Diuron	Water	SPE (C18)	not reported	0.1-1.59 µg/L	SFC-DAD CO_2 acetonitrile (190-360 nm)	Salvatierra-Stamp et al., 2015
6 OCPs	Water	USAEME (isooctane)	not reported	0.8-10 ng/L	GC-MS Q (SIM)	Zhang et al., 2012
40 pesticides	Aqueous samples	SBSE (PDMS)	valve	< 1 ng/L	GC-MS TOF APGC	Pintado-Herrera et al., 2014
9 EDCs (3 herbicides), 19 PPCPs	Water, wastewater irrigated soils	SPE, USE/ SPE (silicagel)	valve	water: 0.15-14.08 ng/L; soil: 0.06-10.64 ng/g	RRLC-MS/MS, QqQ, ESI	Chen et al., 2010
58 EDCs (16 pesticides	Wastewater	SPE (OASIS HLB)/SPE (florisil)	PTV, LVI	0.2-735 ng/L	GC-MS Q (SIM) EI	Bizkarguenaga et al., 2012
25 OCPs	River and wastewater	SBSE (PDMS)	splitless	< 10 ng/L	GC-MS/MS QqQ (MRM) EI	Chary et al., 2012
7 OCPs	Rain water	SPME (sol-gel graphene)	spliless	0.19-18.3 ng/L	GC-MS Q (SIM)	Ke et al., 2013
5 pesticides	Well water	SPME (polythiophen)	splitless	0.02-0.64 ng/mL	GC-ECD	Pelit and Dizdas, 2013
7 pesticides	Well water	SFODME (2-dodecanol)	splitless	0.0002-0.043ng/mL	GC-ECD	Pelit et al., 2014
12 pesticides	Spring water	SPE (Oasis HLB)	splitless	0.0004-0.29 µg/L (LOQ)	GC-MS Q (SIM) EI	Bono-Blay et al. , 2012

Table 3. (Continued)

Analytes	Matrix	Extraction/ clean-up	Injection technique	LOD	Separation and detection technique	References
97 organic contaminants (17 OCPs, 14 OPP, 7 PP)	Coastal sediments	ASE(DCM)/in cell clean up (alumina)	splitless	< 1 ng/g	GC-MS/MS QqQ (MRM) EI	Pintado-Herrera et al., 2016
8 pesticides	Environmental samples	LLE (n-hexane:DCM (1:9))/SPE (florisil)	splitless	0.94-8 ng/g	GC-MS Q (FS) EI	Sichilongo and Banda, 2013
69 EDCs (2 pesticides, lindane, cypermethrine)	Indoor air	ASE (DCM: methanol 1:2)	valve	< 600 pg/m^3	12 EDCs: LC-MS/MS QqQ 27 EDCs: GC-MS Q 30 EDCs: GC-MS/MS QqQ EI	Laborie et al., 2016
Biological samples						
Dicofol, DDTs	Human milk	LLE, GPC	splitless	0.1–0.2 ng/g	GC-MS (SIM), EI	Fujii et al, 2011
34 EDCs (24 pesticides, OCPs, OPPs)	Human fetal and newborn tissues	SPE (C18)	splitless	0.16-1.6 ng/g	GC-MS Q (SIM) EI	Cappiello et al., 2014
3 OCPs	Human serum	SPE (alumina)/SPE (silica)	not reported	1 ppb ng/g	GC-ECD	Arrebola et al., 2012
18 OCPs	Human serum	SPME (polyacrylate)	split	0.01-1 ng/mL	GC-MS Q (SIM) EI	Kim et al., 2013
Chlorpyrifos and metabolites	Urine samples	SFODME (2-dodecanol)	splitless	3.8-4.8 ng/L	GC-MS Q (SIM) EI	Pelit and Yengin, 2014

Analytes	Matrix	Extraction/ clean-up	Injection technique	LOD	Separation and detection technique	References
Cis-permethrin, trans-permethrin and metabolites (insecticides)	Biological rat matrices	LLE (toluene)	splitless	25-50 ng/mL	GC-MS/MS IT (SRM)	Lestremau et al., 2014
Atrazine and vinclozolin metabolite	Rat serum	SPE (Oasis HLB)	valve	0.008-4 ng/mL	HPLC-MS/MS QTrap (MRM) ESI	Tournier et al., 2015
10 pesticides and metabolites	Fish bile	SPE (Plexa cartridge)/ dSPE (florisil)	splitless	0.04-459 ng/mL	GC-MS Q (SIM) EI	Ros et al., 2015

Abbreviations: APGC – atmospheric pressure gas chromatography, ASE – accelerated solvent extraction, DAD – diode array detector, DCM – dichlormethane, DDT – dichlordiphenyltrichlorethane, DLLME – dispersive liquid liquid microextraction, dSPE – dispersive solid phase extraction, DVB – divinylbenzene, ECD – electron capture detector, EDCs – endocrine disrupting chemicals, EDPs – endocrine disrupting pesticides, EI – electron ionization, ESI – electrospray ionization, EtAc – ethylacetate, FS – full scan, GC – gas chromatorgraphy, GPC – gel permeation chromatography, IT – ion trap, LIT – linear ion trap, LLE – liquid-liquid extraction, LOD – limit of detection, LVI – large volume injection, MCX – mixed-mode, reversed-phase/strong cation-exchange polymer (Oasis MCX), MeCN – acetonitrile, MNP – magnetic nanoparticles, MRM – multiple reaction monitoring, MS – mass spectrometry, MS/MS – tandem mass spectrometry, MSPD – matrix solid phase dispersion, NCI – negative chemical ionization, OCPs – organochlorine pesticides, OPPs – organophosphorous pesticides, PDMS – polydimethylsiloxane, PP – pyrethroides, PPCP – pharmaceuticals and personal care products, PSA – primary secondary amine, PTV – programmed-temperature vaporization (injector), Q – quadrupole, QqQ – triple quadrupole, QuEChERS – quick, easy, cheap, effective, rugged and safe, RRLC - rapid resolution liquid chromatography, SBSE – stir bar sorptive extraction, SDME – single drop microextraction, SFC – supercritical fluid chromatography, SFODME – solidification of floating organic drop microextraction, SIM – selected ion monitoring, SPE – solid phase extraction, SPME – solid phase microextraction, SRM – selected reaction monitoring, SVV – solvent vent valve, TOF – time-of-flight, USAEME – ultrasound-assisted emulsification microextraction, USE – ultrasonic extraction.

For GC-MS multiresidue analysis, electron ionization (EI) and chemical ionization (CI) can be employed. EI is the ionization technique of the first choice; however, quite often the lack of information due to the extensive fragmentation of the molecular ions requires the use a softer ionization mode, PCI (positive CI) or NCI (negative CI), which produces fewer ion fragments (Béguin et al., 2006). NCI results in "cleaner" chromatograms with reduced interferences from ions derived from the sample matrix; better signal-to-noise ratio; higher sensitivity and selectivity. Detection limits for many pesticides/analytes are two orders of magnitude lower than the corresponding EI-MS or PCI-MS methods (Húšková et al., 2010*b*). NCI-MS was coupled with fast GC for separation and detection of EDPs in fruit and vegetables, and for the majority of pesticides the lowest calibration levels were 0.01 to 0.05 μg/kg for fast GC-NCI-MS and 1 μg/kg for fast GC-EI-MS (Húšková et al., 2009). There is scarce literature information about the NCI MS/MS. SANTE document, 2015 recommends retention time plus two transitions and the correct ratios between them for satisfactory analyte identification and confirmation. Belmonte Valles et al., 2012 studied the NCI-MS/MS, and found that for many compounds, such as chlorpyrifos methyl, tetraconazole, pyrifenox, endosulfan alpha, fenpropathrin, boscalid and azoxystrobin only one transition is available.

For LC-MS, atmospheric pressure interfaces (APIs) are the most relevant, namely electrospray ionization (ESI) and atmospheric pressure chemical ionization (APCI), and were selected as the most sensitive modes for ionization of target compounds before MS detection. ESI is soft ionization technique, it is the most important ionization technique for compounds of moderate to high polarity (<1 kDa) (Smyth, 2003). APCI is better suited for smaller and less polar analytes. Atmospheric pressure photo-ionization (APPI), introduced just a few years ago is used for special cases, e. g. PAHs. It did not reach a widespread application.

A new group of ionization techniques has emerged in the last few years. These techniques generate ions from samples under ambient conditions. The most common ambient ionization techniques applied in pesticide residues analysis are the following: desorption electrospray (DESI), direct analysis in real time (DART), and extractive electrospray ionization (EESI).

Within scanning analyzers, quadrupole (Q) analyzer has been most widely utilized for quantification in GC and HPLC. Quadrupole can operate in full scan or selected ion monitoring (SIM) mode. To obtain sufficiently low detection limits, required for regulation purpose, the SIM mode has to be employed. Sensitivity is enhanced by monitoring only a few ions of interest, but the spectral information is lost. It was shown by Kirchner et al., 2005, that the coupling of fast GC utilizing narrow-bore column (with I.D. of 0.15 mm) with a common bench top quadrupole MSD provides efficient separation of 18 studied pesticides from the apple matrix components, sufficient number of data-points acquired for the proper peak shape reconstruction and good repeatability of peak areas in SIM mode and overall system ruggedness after repeated matrix injections.

Application of single quadrupoles in LC-MS has not had much success owing to low sensitivity, high background noise and poor structural information obtained in the soft ionization sources employed. If single quadrupole is not sufficient for the detection of analytes due to the high chemical noise and influence of the matrix, MS detectors can be employed in tandem mass spectrometric arrangement (MS/MS, MS^2) in order to offer higher selectivity and sensitivity of detection (Maggioni et al., 2013; Chen et al., 2010; Laborie et al., 2016). The widespread arrangement of tandem MS in space is triple quadrupole (QqQ) MS, which is able to isolate the molecular ion of the compound of interest in the first stage of the mass analyzer and obtain selective precursor-product ion transitions when operated in selected reaction monitoring (SRM) or multiple reaction monitoring (MRM) mode. The most intense fragment ion from the precursor ion is used for quantification. A less sensitive secondary transition is used as the second criterion for confirmation purposes (Sosa-Ferrera et al., 2013). In spite of the large number of analytical methods based on the use of QqQ, the quantification of contaminants can present some drawbacks: for reliable quantification, two SRM transitions are required and some analytes present only one transition; some transitions are unspecific; SRM are typically limited to 100 − 150 target analytes, depending on the chromatographic separation, resulting in a loss of sensitivity (Farré and Barceló, 2013). In addition, this approach results in a target analysis workflow where the set-up of the acquisition method is extensive and time-consuming when a few hundred pesticide residues are analyzed simultaneously in a single run. Another drawback of this approach is its inability to detect target compounds which have not been previously optimized; or to perform retrospective data analysis (Gómez-Ramos et al., 2013). In combination with GC, MS/MS employing QqQ system is routinely used at present (Fernandes et al., 2012; Munaretto et al., 2013; Mansilha et al., 2010; Chary et al., 2012; Pintado-Herrera et al., 2016). For the case of multianalyte complex indoor samples, three arrangements were applied, polybrominateddiphenylethers, polychlorinatedbiphenyls, chlorobenzenes, and lindane were analyzed by GC-MS/MS phthalates, synthetic musks, cypermethrin and polyaromatic hydrocarbons were determined by GC-MS and parabens, bisphenols, triclosan, hexabromocyclododecane, alkylphenols and alkylphenol ethoxylates were quantified by LC-MS/MS (Laborie et al., 2016).

Ion trap (IT) analyzers offer routine GC–MS/(MS^n) capability without the high cost or added complexity of multisector instrument. Similar to Q-MS, IT-MS detection also offers limited number of ions (consequently number of analytes), which can be recorded in one time window. In MS^2 arrangements, the precursor ions are first selected from the injection of a high-concentration standard. Then, the fragmentation pattern is determined by applying various energies. Product ions are chosen from the ions that provided the maximum intensities for the various fragments. Lestremau et al., 2014 utilized GC-MS/MS with IT for quantification of selected pyrethroids. The selectivity obtained for complex matrices such as rat blood and organs allowed to eliminate purification step for most of the matrices investigated. Linear IT (Lin IT) is often considered as a version of the threedimensional (3D) trap (Douglas et al., 2005) and it is a trapping device of greater capacity than the 3D trap. Lin IT is often employed as the first device in a tandem MS, especially if the spectrometer is being used with ESI. Li et al., 2013 developed the method for determination of 19 triazine pesticides and degradation products in processed cereal samples employing Lin IT in MS^3 arrangement. It was shown that the detection limits of MS^3 in comparison with MS^2 were lower because of less baseline noise. Tandem QTrap in MRM mode was utilized for multifamily analytes including hormones and vinclozolin metabolites and atrazine in rat serum (Tournier et al., 2015).

Targeted approach limits the number of compounds that can be investigated. Mass spectrometers can operate in scan mode and thus allow the untargeted approach to overcome the restrictions encountered with target analysis. Time-of-flight (TOF) analyzers are the choice for non-targeted screening, because spectral information is saved for the whole selected mass range. TOF allows the simultaneous sampling and analysis of all ions across the whole range. The advantage of the TOF-MS for screening is the ability to examine a data file for theoretically unlimited number of pesticides (e.g., hundreds of compounds) with high sensitivity within one run (Garcia-Reyes et al., 2007). TOF is considered to be the detector of choice primarily for identification purposes. TOF data are generated independently of the list of compounds screened, which enables retrospective study of the data without the need to re-analyse the samples. The combination QTOF-MS/MS for pesticide residues analysis in fruits and vegetables provides a reliable tool for identification of pesticide transformation products and metabolites, fragmentation pattern together with accurate-mass measurement of product ions, thus providing identification and confirmation with a single system (Ferrer et al., 2012).

Fourier transform ion cyclotron resonance (FT-ICR) is one of the most sensitive ion-detection methods with mass accuracy of <1 ppm, but the high cost of instrumentation currently restricts its application (Farré and Barceló, 2013). The enormous cost of FT-ICR prompted work on new entry in the field of analyzers, Orbitrap. Orbitrap mass analyzer was launched in 2005 and it provides an interesting alternative to other commonly employed high-resolution (HR) analyzers; however, it suffers from a slower data acquisition rate than QTOF instruments. On the other hand, it provides outstanding mass accuracy and mass resolution. HRMS enables the simultaneous detection of non-target pesticides, giving the advantage to search for transformation products of pesticides in the analysed samples (Páleníková et al., 2015[b]). Currently, there are only limited studies on the use of Orbitrap instruments for the quantitation of pesticides.

MATRIX EFFECTS IN PESTICIDE RESIDUES ANALYSIS

None of the common isolation and/or clean-up techniques is selective enough to avoid the presence of residual sample components in the analytical sample. Matrix-induced chromatographic response enhancement or suppression is a phenomenon that causes overestimation of the analytes concentration, or excessively high recovery results, that arise from higher detector responses observed when a substance is injected in a matrix-modified standard solution (matrix-standard) compared to an injection in pure solvent or matrix-free standard solution (Hajšlová and Čajka, 2006). When a real sample is injected, the matrix components tend to block active sites in the GC injector and column, thus reducing losses of susceptible analytes caused by adsorption or degradation on active sites. Individual analytes in complex matrices such as food or environmental samples may be affected to a different extent by this phenomenon, and changes may occur also from sample to sample within the varieties or sortes. The most common strategies to compensate for the matrix effects in GC-MS include: use of isotopically labelled internal standards; method of standard addition; use of matrix-matched standards – standards with the same matrix composition as the analyzed samples; and use of analyte protectants (APs) (Hrouzková and Matisová, 2012). However, none of these techniques is universal for routine analysis.

The occurrence of matrix effects is also a problem in quantitative analysis by LC-MS. Matrix effects in LC-MS occur because of the interaction of co-eluting compounds with the analytes in the ionization process, producing, in some cases, signal suppression (Gómez-Ramos et al., 2013). Reducing or eliminating matrix effect as reviewed by Kruve et al., 2009 can be achieved through sample dilution, decreasing flow rate via post-column flow splitting, changing the ionization source (to APCI for example), or mode, and improving sample preparation. In ESI the matrix effect is believed to be largely due to competition between different compounds for the surface on the ESI droplets. A sample dilution decreases the number of these competing molecules per microdroplet, the ionization efficiency increases and, thus, so does the analyte signal. Sample dilution as well as smaller sample volume injection decreases the number of competing molecules per drop area and thus also the matrix effect (Kruve et al., 2009). Compensation of matrix effects in LC-MS is more difficult to achieve because the matrix effects depend on the co-elution of individual pesticide with co-extracted matrix components, which vary between different commodities.The use of matrix-matched calibration is, therefore, likely to be less effective compared to GC-MS (SANTE document, 2015).

Matrix-induced effects, such as signal enhancement or suppression, are ordinarily evaluated using the matrix factor (MF). The MF is calculated for each pesticide by comparing the analyte response in matrix-matched solution vs. the pesticide response obtained in pure solvent, according to the following equation (Páleníková et al., 2015[a]):

$$MF = \left(\frac{response \quad in \quad matrix - matched \quad solution}{response \quad in \quad solvent \quad solution} - 1\right) \times \ 100\%$$

Exceeding the MF higher than 20% or smaller than −20% indicates the peak signal enhancing or suppressing due to the matrix effects in the calibration procedure (SANTE document, 2015).

Matrix-matched standards were originally used for investigation of pesticides in food samples, but to avoid and minimize any ambiguity related to the matrix induced enhancement effect, calibration curves for quantification of EDPs in water samples are also generated from matrix-matched standards (Mansilha et al., 2010; Purdešová et al., 2013). Signal enhancement observed for several compounds in wastewater effluents was managed by the use of matrix-matched standards and different injection liners (Chary et al., 2012).

Belmonte Valles et al., 2012 studied the influence of dilution on the matrix effect in GC-MS/MS – spiked extracts from 5 to 500 μg/kg were diluted ten times. It is necessary to notice that NCI and MS/MS approach was used to reach satisfactory LOQ. There are many different components in each variety of tomato and apple, but the slopes of matrix-matched calibration curve between the different varieties of apple and tomato were similar when diluted. This fact allowed one to perform the quantification of real samples using any variety to prepare matrix-matched standards.

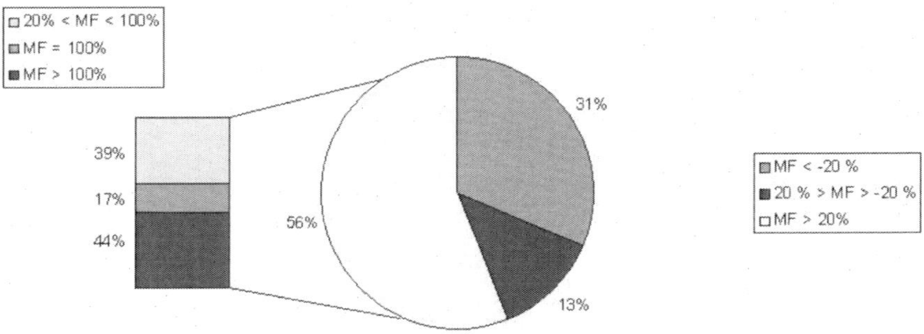

Figure 4. Distribution of pesticides affected/un-affected by matrix effect according to MF values.

Multi-residue method of 32 EDPs belonging to different groups (organochlorines, organophosphates, pyrethroids, dicarboximides, 2,6-dinitroanilines, triazinones, substituted ureas, phthalamides, cyclodienes, triazoles, imidazoles), varying in polarity, volatility and other physicochemical properties from fatty fruit and vegetable matrices based on fast GC with narrow-bore column was developed by our group (Hrouzková, 2016). Pesticide residues analysis in fatty samples is complicated by the co-injected matrix constituents responsible for a severe matrix effect. Pesticides were isolated from pumpkin seed by ASE and were analysed by GC-MS. Matrix-matched standards were used to compensate for the matrix effects and to investigate the difference between responses in matrix-matched standards and standards in solvent at the concentration level of 10 μg/kg, which is considered the lowest MRL value within the investigated combinations of pesticide/matrix. Figure 4 shows the percentage of pesticides affected by matrix effect expressed as the MF value. Only 13% of the pesticides showed small or no matrix effect (MF values between −20 and 20%). Most of the pesticides (56%) exhibited signal enhancement effects with MF values higher than 20%. It can be seen that approximately one third of pesticides exhibited signal suppression. As a conclusion, most of the MFs are out of the acceptable range, indicating significant differences between the matrix-matched standards and the solvent-matched standards responses. Therefore, matrix-matched standard solutions were used to eliminate the matrix effect in real-sample analyses.

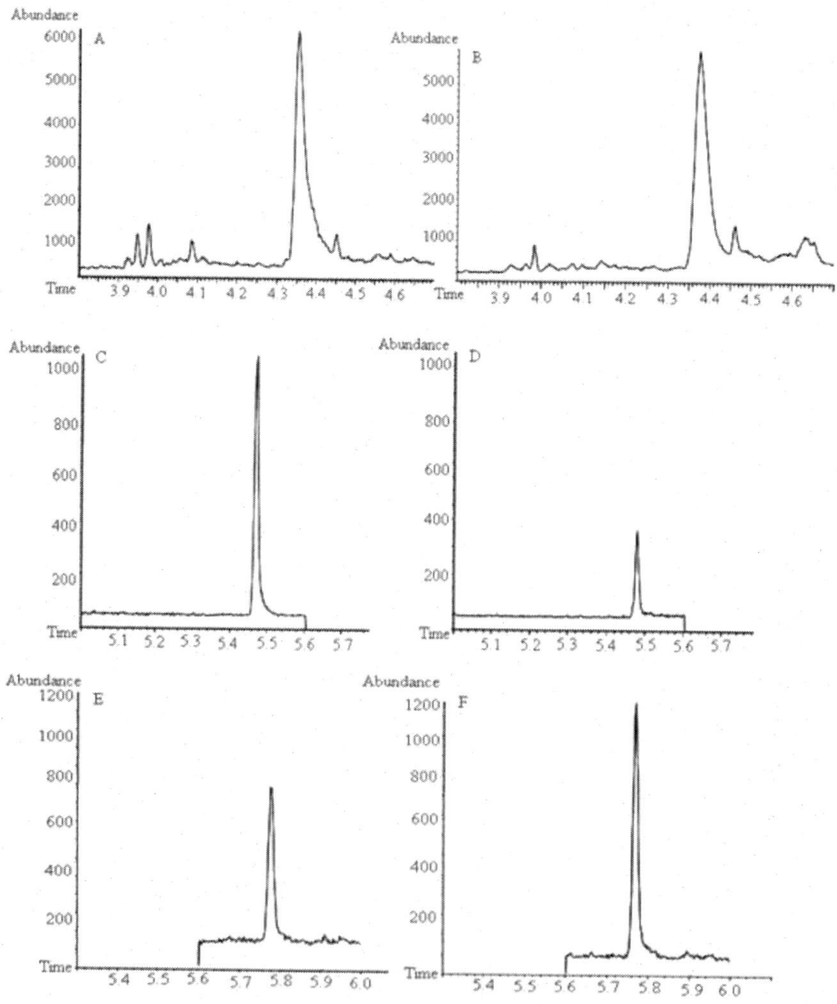

Figure 5. Chromatogram of target ions of selected EDPs: A) o- phenylphenol in matrix-matched standard solution, B) o-phenylphenol in solvent solution, C) trifluralin in matrix-matched standard solution, D) trifluralin in solvent solution, E) hexachlorbenzene in matrix-matched standard solution and F) hexachlorbenzene in solvent solution (high lipid content matrix 10 μg/kg).

For illustration of 3 possibilities of matrix effects, Figure 5 shows the chromatogram of quantifier ions of selected EDPs in matrix-matched standard (prepared from pesticide-free pumpkin seeds) solution vs. solvent solution. Figures 5 A, B show no matrix effect, Figures 5 C, D show signal enhancement, and Figures 5 E, F illustrate signal suppression.

Table 4A. Occurrence of EDPs reported in real-life food samples

Compounds	EDC Cat.	Matrix	Conc. μg/kg	Ref.
Endosulfan sulfate	2, TEDX	Green bean	5.4	Belmonte Valles, 2012
Atrazine and metabolites	1, TEDX	Orange, cereal samples, drinking water	0.007-0.356; 0.05-13.09 ng/L	Andraščíková, 2013; Li, 2013; Maggioni, 2013
Dimethoate	2, TEDX	Orange	27-529	Andraščíková, 2013
Vinclozolin	1, EPA, TEDX		1	
HCB	1, TEDX	Strawberry jam	1.0-1.3	Fernandes, 2012
Aldrin	2, TEDX		0.4-0.5	
Endrin	2, TEDX		2.0	
Lindane	1, EPA, TEDX	Strawberry jam, strawberries	1.0-27.1	
β-Endosulfan	2, TEDX	Strawberries	7.1	
Simazine	2, TEDX	Cereal samples, drinking water	0.805; 0.09-1.59 ng/L	Li, 2013; Maggioni, 2013
Cyromazine	EPA	Cereal samples	0.93-0.987	Li, 2013
p,p´-DDT	1, TEDX	Fish	0.50-9.13	Santhi, 2012
Linuron	1, TEDX	Drinking water	0.11-0.44 ng/L	Maggioni., 2013

Occurrence of EDPs in Real-Life Samples

Validated analytical methods for determination of EDPs at low concentration levels are essential for surveillance programs that are established with the goal of minimizing the hazards and the risks to the health. Various EDPs proven or suspected to disrupt endocrine or immune systems have been detected in food or environmental samples. For illustration, a selection of recent findings focused mainly on high-volume produced chemicals and EDCs belonging to proven or suspected EDPs published in scientific literature since 2012 is shown in Table 4. Table 4A is devoted to food samples, while Table 4B summarizes findings in the environmental samples. These pesticides are listed in category 1 or 2 according to the classification system by EC (EC, 2011), or in US EPA screening program (US EPA), or on endocrine disruption fact sheet created by TEDX. TEDEX –*The endocrine disruption exchange*, is a nonprofit organization dedicated to compiling and disseminating the scientific evidence on the health and environmental problems caused by EDCs, (Colborn, TEDX).

Table 4B. Real findings of EDPs reported in environmental samples

Compounds	EDC (Cat.)	Matrix	Conc. ng/L	Ref.
Dieldrin	2, TEDX	River water	80	Zhang, 2012
p,p´-DDT	1, TEDX	Wastewater	9.3-1057	Bizkarguenaga, 2012; Ke, 2013
Alachlor	1, EPA, TEDX	River, wastewater	11.2-52.2	Bizkarguenaga, 2012; Chary, 2012
Lindane	1, EPA, TEDX	Wastewater, sediments, river	7.6-1900; 1.3-1.6 ng/g	
Atrazine and metabolites	1, TEDX	Wastewater, well water	2-79	Bizkarguenaga, 2012; Bono-Blay, 2012
Simazine	2, TEDX		1-18	
Vinclozolin	1, EPA, TEDX	River, wastewater	1.7-12.8	Chary, 2012
Metolachlor	EPA, TEDX		1.2-53.2	
Endosulfan sulphate	2, TEDX		3.1-30.2	
β-Endosulfan	2, TEDX	River, wastewater, soil	0.2-21.6; 1.9-7.4 μg/kg	Chary, 2012, Fernandes, 2012
HCB	1, TEDX	Rainwater	62.7	Ke, 2013
Trans-chlordane	1, TEDX		79.6	
Aldrin	2, TEDX	Well water, sediments, soil	50; 0.97-20.4 ng/g	Pelit et al., 2014, Pintado-Herrera, 2016; Fernandes, 2012
Iprodione	2, TEDX	Well water	80-260	Pelit et al., 2014
Endrin	2, TEDX	Soil	0.2-5.2 μg/kg	Fernandes, 2012

Notes: DDT– dichlorodiphenyltrichloroethane, HCB- hexachlorbenzene.
TEDX – http://www.endocrinedisruption.org/.

The most frequently identified EDPs are OCPs. OCPs have been used in agriculture for a long time to control a variety of crop diseases and pests, and also to control malaria spreading. Many of them have been banned or restricted. Due to their persistent nature, chemical stability and lipophilic character, they can be present in environmental components, in food or bioaccumulated in lipid tissue of organizms. Hexachlorobenzene (HCB) is known as a by-product of chlorination and its occurrence in water samples (Ke et al., 2013) is due to industrial contamination.

Since the OCPs have been withdrawn from registered use, OPPs are currently the largest and the most versatile class in use due to better diseases and insect pests control and faster degradation in the environment (Yang et al., 2015). Dimethoate is an insecticide, which was identified in orange samples. In three samples, the concentration of dimethoate exceeded the MRLs established by the EU (Andraščíková et al., 2013).

Iprodion is a dicarboximide fungicide widely used in fruits and grapevines. There is a potential for it to transfer from soil to water with resulting risk to human health. Well water samples were contaminated with iprodione in different areas of vineyards. The concentration levels found for iprodione were above the limit established by IUPAC Technical Reports – 0.01 μg/kg for drinking water (Pelit et al., 2014; Hamilton et al., 2003).

Triazine pesticides have been widely used to control annual and perennial grass in various crops. The most widely used triazines have been listed in the national and EU coordinated monitoring program for the routine monitoring and consumer risk assessment (European Comission, 2006). Atrazine and its metabolites were identified in 18 of the 35 drinking water samples (Maggioni et al., 2013) and in 5 orange samples (Andraščíková et al., 2013). This herbicide was banned several years ago, but unacceptable values still found may be due to its widespread use in agriculture all over the world and because it is a presistent organic pollutant (Andraščíková et al., 2013).

Simultaneous screening of phthalates, alkylphenols, bisphenol A and herbicides in 131 groundwater samples was done. Knowledge on the occurrence of contaminants in source waters intended for bottling is important for sanitary and regulatory purposes. Among herbicides simazine, atrazine and its metabolites were detected in seven samples. It was shown that the impact of herbicides and other EDCs in 131 source waters is minimal (Bono-Blay et al., 2012).

The degradation products and metabolites of EDPs are often more toxic than parent types. Chlorpyrifos is an insecticide suspected to be EDC and its metabolite - chlorpyrifos–oxon is more toxic than its parent molecule. It was shown, that sub ppb levels of oxon metabolite are relevant and necessary to screen in clinical samples (Pelit and Yengin, 2014).

As it was shown in this subchapter, numerous EDPs findings were reported in scientific literature. Although the concentration levels are low, ordinarily below the permitted limit, in the case of EDPs even tiny amounts could be potentially harmful from the long-term point of view. Effects of EDPs may happen at extremely low doses; they may affect multiple signaling systems that control function and development; they may be subtle, long-term and/or delayed; and through parental exposure they may even affect subsequent generations (Vandenberg et al., 2012). Another important point is the synergic effect among various EDCs. Numerous cases of real-life findings of EDPs in the real samples indicate that there is a need for establishing monitoring programs in order to assure the safety of the food products and the environment.

Conclusion

Many pesticides could disrupt normal function of endocrine and immune systems in living organizms. EDPs could enter food or environment through various pathways and harm the hormone function. Food and environment are complex samples and it is necessary to monitor the EDPs occurrence to protect human health as wel as all compartments of the environment.

EDCs belong to the most widely analyzed chemicals. Analysis of complex samples typically requires an extraction step prior to the separation and the detection. For extraction step, several techniques have been developed. The frequent need of the purification steps was demonstrated. Current trends in sample preparation of EDPs include usage of different techniques with the aim to simplify the methodology and lower solvent consumption. An extensive development has been performed to obtain greener methodologies, such as different kinds of microextractions. Preparation of new sorbents is an important way to enhance method selectivity.

Multiclass multiresidue methods require effective separation either by gas or liquid high-resolution chromatography, with the aim to separate components as fast as possible. Fast separation results in decreased total analysis time and increased sample throughput.

For detection of EDPs, mass-spectrometric detection techniques show excellent sensitivity and precision. The application of advanced MS analyzers such as QqQ, TOF, qTOF, LIT, and Orbitrap to food and environmental sample analysis has broaden the range of analytes that can be identified and quantified at low concentration levels.

The challenges for the future research include identification of new emerging endocrine disruptors, their metabolites and transformation products, increasing the reliability of identification, lowering limits of detection/quantification and development of "green chemistry" methodologies that will reduce the impact on the environment.

ACKNOWLEDGMENTS

This work was supported by the Scientific Grant Agency of the Ministry of Education of the Slovak Republic and the Academy of Sciences (VEGA, project No. 1/0503/14) and by the STU Grant scheme for Support of Excellent Teams of Young Researchers.

REFERENCES

Abdel-Rehim, M., Altun, Z.& Blomberg, L. (2004). Microextraction in packed syringe (MEPS) for liquid and gas chromatographic applications. Part II—Determination of ropivacaine and its metabolites in human plasma samples using MEPS with liquid chromatography/tandem mass spectrometry. *J. Mass Spectrom.*, *39*, 1488-1493.

Abdel-Rehim, M. (2011). Microextraction by packed sorbent (MEPS): a tutorial. *Anal. Chim. Acta*, *701*, 119-128.

Ahmad, W., Al-Sibaai, A. A., Bashammakh, A. S., Alwael, H. & El-Shahaw, M. S. (2015). Recent advances in dispersive liquid-liquid microextraction for pesticide analysis. *TrAC Trends in Anal. Chem.*, *72*, 181-92.

Almeida, C., Serôdio, P., Florencio, M. H.& Nogueira, J. M. F. (2007). New strategies to screen for endocrine-disrupting chemicals in the Portuguese marine environment utilizing large volume injection-capillary gas chromatography-mass spectrometry combined with retention time locking libraries (LVI-GC-MS-RTL). *Anal. Bioanal. Chem.*, *87*, 2569-2583.

An, Y. R., Joo, Y. K., Kim, S. J., Yu, S. Y., Ahn, J. J., Hong, J. Y.& Hwang, S. Y. (2014). Functional analysis of endocrine disruptor pesticides affected transcriptome and microRNA regulation in human hepatoma cell line. *Mol. Cell. Toxicol.*, *10*, 393-400.

Anastassiades, M., Lehotay, S. J., Štajnbaher, D. & Schenck, F. J. (2003). Fast and easy multiresidue method employing acetonitrile extraction/ partitioning and "dispersive solid-phase extraction" for the determination of pesticide residues in produce. *J. AOAC Int.*, *86*, 412-431.

Andraščíková, M., Hrouzková, S. & Cunha, S. C. (2013). Combination of QuEChERS and DLLME for GC-MS determination of pesticide residues in orange samples. *Food Add. Contam.*, *30*, 286-297.

Andraščíková, M.& Hrouzková, S. (2013). A comparative study of three modifications of QuEChERS method for determination of endocrine disrupting pesticide residues in lemon matrices by fast GC-MS. *Anal. Methods*, *5*, 1374-1384.

Andraščíková, M., Matisová, E.& Hrouzková, S. (2015). Liquid phase microextraction techniques as a sample preparation step for analysis of pesticide residues in food. *Sep. & Purif. Revs.*, *44*, 1-18.

Androutsopoulos, V. P., Hernandez, A. F., Liesivuori, J.& Tsatsakis, A. M. (2013). A mechanistic overview of health associated effects of low levels of organochlorine and organophosphorous pesticides. *Toxicology*, *307*, 89-94.

Annamalai, J.& Namasivayam, V. (2015). Endocrine disrupting chemicals in the atmosphere: Their effects on humans and wildlife. *Env. Int.*, *76*, 78-97.

Arrebola, J. P., Cuellar, M., Claure, E., Quevedo, M., Antelo, S. R., Mutch, E., Ramirez, E., Fernandez, M. F., Olea, N.& Mercado, L. A. (2012). Concentrations of organochlorine pesticides and polychlorinated biphenyls in human serum and adipose tissue from Bolivia. *Env. Res.*, *112*, 40-47.

Arthur, C.& Pawliszyn, J. (1990). Solid phase microextraction with thermal desorption using fused silica optical fibers. *Anal. Chem.*, *62*, 2145-2148.

Asensio-Ramos, M., Ravelo-Pérez, L. M., González-Curbelo, M. Á. & Hernández-Borges, J. (2011). Liquid phase microextraction applications in food analysis. *J. Chromatogr. A.*, *1218*, 7415-7437.

Baltussen, E., David, F., Sandra, P., Janssen, H. G.& Cramers, C. (1999). Automated sorptive extraction-thermal desorption-gas chromatography-mass spectrometry analysis.Determination of phenols in water samples.*J. Microcol. Sep.*, *11*, 471-474.

Barker, S. A. (2000). Matrix solid-phase dispersion. *J. Chromatogr. A*, *885*, 115-127.

Barrida-Pereira, M., Serodio, P., Gonzalez-Castro, M. J.& Nogueira, J. M. F. (2010). Determination of organochlorine pesticides in vegetable matrices by stir bar sorptive extraction with liquid desorption and large volume injection-gas chromatography-mass spectrometry towards compliance with European Union directives. *J. Chromatogr. A*, *1217*, 119-126.

Béguin, S., Jadas-Hécart, A., Tabet, J. C.& Communal, P. (2006). Protocols for optimizing MS/MS parameters with an ion trap GC-MS instrument. *J. Mass Spectrom.*, *41*, 1304-1314.

Belmonte Valles, N., Retamal, M., Mezcua, M.& Fernández-Alba, A. R. (2012). A sensitive and selective method for the determination of selected pesticides in fruit by gas chromatography/mass spectrometry with negative chemical ionization. *J. Chromatogr. A*, *1264*, 110-116.

Beyer, A. & Biziuk, M. (2008). Applications of sample preparation techniques in the analysis of pesticides and PCBs in food. *Food Chem.*, *108*, 669-680.

Bicho, R. C., Amaral, M. J., Faustino, A. M. R., Power, D. M., Rema, A., Carretero, M. A., Soares, A. M. V. M. & Mann, R. M. (2013). Thyroid disruption in the lizard Podarcis bocagei exposed to a mixture of herbicides: a field study. *Ecotoxicology*, *22*, 156-165.

Bidari, A., Reza Ganjali, M., Norouzi, P., Reza Milani Hosseini, M &, Assadi, Y. (2011). Sample preparation method for the analysis of some organophosphorus pesticides residues in tomato by ultrasound-assisted solvent extraction followed by dispersive liquid–liquid microextraction. *Food Chem.*,*126*, 1840-1844.

Bizkarguenaga, E., Ros, O., Iparraguirre, A., Navarro, P., Vallejo, A., Usobiaga, A.& Zuloaga, O. (2012). Solid-phase extraction combined with large volume injection-programmable temperature vaporization–gas chromatography–mass spectrometry for the multiresidue determination of priority and emerging organic pollutants in wastewater. *J. Chromatogr. A*, *1247*, 104-117.

Bono-Blay, F., Guart, A., Fuente, B., Pedemonte, M., Pastor, M. C., Borrell, A. & Lacorte, S. (2012). Survey of phthalates, alkylphenols, bisphenol A and herbicides in Spanish source waters intended for bottling. *Environ. Sci. Pollut. Res.*, *19*, 3339-3349.

Buszewski, B. & Szultka, M. (2012). Past, Present, and Future of Solid Phase Extraction: A Review. *Critical Revs. Anal. Chem.*, *42*, 198-213.

Camel, V. (2001). Recent extraction techniques for solid matrices – supercritical fluid extraction, pressurized fluid extraction and microwave-assisted extraction: their potential and pitfalls. *Analyst*, *126*, 1182-1193.

Camino-Sánchez, F. J., Rodríguez-Gómez, R., Zafra-Gómez, A., Santos-Fandila, A.& Vílchez, J. L. (2014). Stir bar sorptive extraction: Recent applications, limitations and future trends. *Talanta*, *130*, 388-399.

Cappiello, A., Famiglini, G., Palma, P., Termopoli, V., Lavezzi, A. M. & Matturri, L. (2014). Determination of selected endocrine disrupting compounds in human fetal and newborn tissues by GC-MS. *Anal. Bioanal. Chem.*, *406*, 2779-2788.

Capriotti, A. L., Cavaliere, Ch., Giansanti, P., Gubbiotti, R., Samperi, R. & Laganà, A. (2010). Recent developments in matrix solid-phase dispersion extraction. *J. Chromatogr. A*, *1217*, 2521-2532.

Chang, H. S., Choo, K. H., Lee, B.& Choi, S. J. (2009). The methods of identification, analysis, and removal of endocrine disrupting compounds (EDCs) in water. *J. Hazard. Mater.*, *172*, 1-12.

Chary, N. S., Herrera, S., Gómez, M. J.& Fernández-Alba, A. R. (2012). Parts per trillion level determination of endocrine-disrupting chlorinated compounds in river water and wastewater effluent by stir-bar-sorptive extraction followed by gas chromatography–triple quadrupole mass spectrometry. *Anal. Bioanal. Chem.*, *404*, 1993-2006.

Chen, F., Ying, G. G., Yang, J. F., Zhao, J. L.& Wang, L. (2010). Rapid resolution liquid chromatography-tandem mass spectrometry method for the determination of endocrine disrupting chemicals (EDCs), pharmaceuticals and personal care products (PPCPs) in wastewater irrigated soils. *J. Environ. Sci. Health B*, *45*, 682-693.

Christen, V., Crettaz, P.& Fent, K. (2014). Additive and synergistic antiandrogenic activities of mixtures of azol fungicides and vinclozolin. *Tox. App. Pharm.*, *279*, 455-466.

Cock, M.& Bor, M. (2014). Obesogenic effects of endocrine disruptors, what do we know from animal and human studies? *Env. Int.*, *70*, 15-24.

Colborn, T., vom Saal, F. S.& Soto, A. M. (1993). Developmental effects of endocrine disrupting chemicals in wildlife and humans. *Environ. Health Perspect.*, *101*, 378-384.

Colborn, T. (2006). A Case for Revisiting the Safety of Pesticides: A Closer Look at Neurodevelopment. *Environ. Health Perspect.*, *114*, 10-17.

Colborn, T. TEDX- The endocrine disruption exchange, endocrine disruption fact sheet. Aviable online: http://www.endocrinedisruption.com/pesticides/introduction (accessed on February 12, 2016).

Comerton, A. M., Andrews, R. C.& Bagley, D. M. (2009). Practical overview of analytical methods for endocrine-disrupting compounds, pharmaceuticals and personal care products in water and wastewater. *Philos. Trans. R. Soc. A*, *367*, 3923-3939.

David, F. (2015). Application of Stir-Bar Sorptive Extraction in Food Analysis. In: Comprehensive Sampling and Sample Preparation. Analytical Techniques for Scientists, Vol. 4 – Extraction techniques and Applications: Food and Beverage, Pawliszyn, J. (Ed.) 2012, pp. 473–493, Elsevier.

Diamanti-Kandarakis, E., Bourguignon, J. P., Giudice, L. C., Hauser, R., Prins, G. S., Soto, A. M., Zoeller, R. T.& Gore, A. C. (2009). Endocrine-disrupting chemicals: An endocrine society scientific statement. *Endocr. Rev.*, *30*, 293-342.

Douglas, D. J., Frank, A. J. & Mao, D. (2005). Linear ion traps in mass spectrometry. *Mass Spectrom. Revs.*, *24*, 1-29.

European Comission, (2006). Monitoring of pesticide residues in products of plant oringin in the European Union, Iceland and Liechtenstein, Norway.

EC document, (2011). 4th Report on the implementation of the "Community Strategy for Endocrine Disrupters" a range of substances suspected of interfering with the hormone systems of humans and wildlife (COM (1999) 706). Available online:http://ec.europa.eu/environment/chemicals/endocrine/pdf/sec_2011_1001.pdf (accessed on 20 February 2016).

EC document, (2014). Roadmap. Defining criteria for identifying Endocrine Disruptors in the context of the implementation of the Plant Protection Product Regulation and Biocidal Products Regulation. Available online:http://ec.europa.eu/smart-regulation/impact/planned_ia/docs/2014_env_009/endocrine_disruptors_en.pdf (accessed on 20 February 2016).

EC materials, (2016). Available online: http://ec.europa.eu/environment/chemicals/endocrine/strategy/substances_en.htm (accessed on 7 March 2016).

European Workshop, (1996). European workshop on the impact of endocrine disrupters on human health and wildlife. Report of Proceedings. No. 17549, pp. 125.

Ewence, A., Brescia, S., Johnson, I.& Rumsby, P. C. (2015). An approach to the identification and regulation of endocrine disrupting pesticides. *Food and Chem. Toxicol.*, *78*, 214-220.

Farré, M.& Barceló, D. (2013). Analysis of emerging contaminants in food. *Trends in Anal. Chem.*, *43*, 240-253.

Fekete, S., Schappler, J., Veuthe, J. L.& Guillarme, D. (2014). Current and future trends in UHPLC. *Trends in Anal. Chem.*, *63*, 2-13.

Feng, L., Yang, G., Zhu, L., Xu, J., Xu, X.& Chen, Y. (2016). Distribution and risk assessment of endocrine-disrupting pesticides in drinking water sources from agricultural watershed. *Water Air Soil Pollut.*, *227*, 23-32.

Fernandes, V. C., Subramanian, V., Mateus, N., Domingues, V. F.& Delerue-Matos, C. (2012). The development and optimization of a modified single-drop microextraction method for organochlorine pesticides determination by gas chromatography-tandem mass spectrometry. *Microchim. Acta, 178*, 195-202.

Ferrer, C., Malato, O., Agüera, A. & Fernandez-Alba, A. R. (2012). Application of HPLC–TOF-MS and HPLC–QTOF-MS/MS for Pesticide Residues Analysis in Fruit and Vegetable Matrices. *Comprehens. Anal. Chem.*, *58*, 1-60.

Frye, C. A., Bo, E., Calamandrei, G., Calzá, L., Dessi-Fulgheri, F., Fernández, M., Fusani, L., Kah, O., Kajta, M., Le Page, Y., Patisaul, H. B., Venerosi, A., Wojtowicz, A. K. & Panzica, G. C. (2011). Endocrine disrupters: A review of some sources, effects, and mechanisms of actions on behaviour and neuroendocrine systems. *J. Neuroendocrinol.*, *24*, 144-159.

Fujii, Y., Haraguchi, K., Harada, K. H., Hitomi, T., Inoue, K., Itoh, Y., Watanabe, T., Takenaka, K., Uehara, S.& Yang, H. R. (2011). Detection of dicofol and related pesticides in human breast milk from China, Korea and Japan. *Chemosphere*, *82*, 25-31.

Garcia-Reyes, J. F., Hernando, M. D., Molina-Diaz, A. & Fernandez-Alba, A. R. (2007). Comprehensive screening of target, non target and unknown pesticides in food by LC-TOF-MS. *Trends in Anal. Chem.*, *26*, 828-841.

Ghisari, M., Long, M., Tabbo, A. & Bonefeld-Jørgensen, E. C. (2015). Effects of currently used pesticides and their mixtures on the function of thyroid hormone and aryl hydrocarbon receptor in cell culture. *Tox. App. Pharm.*, *284*, 292-303.

González-Curbelo, M. Á., Hernández-Borges, J., Borges-Miquel, T. M.& Miguel Ángel Rodríguez-Delgado, M. Á. (2013). Determination of organophosphorus pesticides and metabolites in cereal-based baby foods and wheat flour by means of ultrasound-assisted extraction and hollow-fiber liquid-phase microextraction prior to gas chromatography with nitrogen phosphorus detection. *J. Chromatogr. A*, *1313*, 166-174.

González-Curbelo, M. Á., Socas-Rodríguez, B., Herrera-Herrera, A. V., González-Sálamo, J., Hernández-Borges, J.& Rodríguez-Delgado, M. Á. (2015). Evolution and applications of the QuChERS method, *Trends in Anal. Chem.*, *71*, 169-185.

Gómez-Ramos, M. M., Ferrer, C., Malato, O., Agüera, A.& Fernández-Alba, A. R. (2013). Liquid chromatography-high-resolution mass spectrometry for pesticide residue analysis in fruit and vegetables: Screening and quantitative studies. *J. Chromatogr. A*, *1287*, 24-37.

Hajšlová, J. & Čajka, T. (2006). *Gaschromatography-mass spectrometry*, (GC-MS). In: Food Toxicants Analysis. Picó, Y. (Ed.), ISBN: 978-0-444-52843-8, Elsevier, Oxford, UK, pp. 419-473.

Hamilton, D. J., Ambrus, A., Dieterle, R. M., Felsot, A. S., Harris, C. A., Holland, P. T. & Katayama, A. (2003). Regulatory limits for pesticide residues in water, IUPAC Technical report. *Pure Appl. Chem.*, *75*, 1123.

Hercegová, A., Dömötörová, M., Hrouzková, S.& Matisová, E. (2007[a]). Study on pesticide residues in apples, apple-based baby food, and their behaviour during processing using fast GC–MS multiresidue analysis. *Internat. J. Environ. Anal. Chem.*, *87*, 957-969.

Hercegová, A., Dömötörová, M.& Matisová, E. (2007[b]). Sample preparation methods in the analysis of pesticide residues in baby food with subsequent chromatographic determination. *J. Chromatogr. A*, *1153*, 54-73.

Hercegová, A., Húšková, R. & Matisová, E. (2010). Evaluation of different calibration approaches in pesticide residues analysis in non-fatty food using fast GC-MS. *Internat. J. Environ. Anal. Chem. 90*, 188-204.

Herrero, M., Mendiola, J. A., Cifuentes, A. & Ibáñez, E. (2010). Supercritical fluid extraction: Recent advances and applications. *J. Chromatogr. A.*, *1217*, 2495-2511.

Hrouzková, S.& Matisová, E. (2011). Pesticides - Strategies for Pesticides Analysis, Margarita Stoytcheva (Editor), Chap. 6, p. 131. InTech, Croatia.

Hrouzková, S. & Matisová, E. (2012). Endocrine Disrupting Pesticides. In: Pesticide-Advances in Chemical and Botanical Pesticides 1st ed.; Soundararajan R. P., Ed.; InTech, Rijeka, Croatia, pp. 99-126.

Hrouzková, S., Matisová, E., Andraščíková, M., Horváth, M., Húšková, R. & Ďurčanská, J. (2012). Survey of low-level endocrine disrupting pesticides in food matrices in Slovakia. *J. Environ. Anal. Chem.*, *92*, 1546-1562.

Hrouzková, S. (2016). *Food Anal. Methods*, in preparation.

Húšková, R., Matisová, E., Hrouzková, S.& Švorc, Ľ. (2009). Analysis of pesticide residues by fast GC in combination with negative chemical ionization mass spectrometry. *J. Chromatogr. A*, *1216*, 6326-6334.

Húšková, R., Matisová, E., Ondreková, S.& Ďurčanská, J. (2010 [a]). Fast GC-MS of endocrine disrupting chemicals.*Internat. J. Environ. Anal. Chem.*, *90*, 173-187.

Húšková, R., Matisová, E.& Hrouzková, S. (2010[b]). Mass spectrometry with negative chemical ionization and its use in GC-MS analysis of organic pollutants. *Chem. Listy*, *104*, 913-920.

Kajta, M.& Wójtowicz, A. K. (2013). Impact of endocrine-disrupting chemicals on neural development and the onset of neurological disorders. *Pharmacol. Rep.*, *65*, 1632-1639.

Kataoka, H., Ishizaki, A., Nonaka, Y.& Saito, K. (2009). Developments and applications of capillary microextraction techniques: a review. *Anal. Chim. Acta*, *655*, 8-29.

Kawaguchi, M., Takatsu, A., Ito, R.& Nakazawa, H. (2013). Applications of stir-bar sorptive extraction to food analysis. *TrAC Trends in Anal. Chem.*, *45*, 280-293.

Ke, Y., Zhu, F., Zeng, F., Luan, T., Su, C.& Ouyang, G. (2013).Preparation of graphene-coated solid-phase microextraction fiber and its application on organochlorine pesticides determination. *J. Chromatogr. A*, *1300*, 187-192.

Khalili Zanjani, M. R., Yamini, Y., Shariati, S.& Jönsson, J. A. (2007).A new liquid-phase microextraction method based on solidification of floating organic drop. *Anal. Chim. Acta*, *585*, 286-293.

Kim, J. Y., Yi, B. R., Go, R. E., Hwang, K., Nam, K. H.& Choi, K. C. (2014). Methoxychlor and triclosan stimulates ovarian cancer growth by regulating cell cycle- and apoptosis-related genes via an estrogen receptor-dependent pathway. *Environ. Toxicol. Phar.*, *37*, 1264-1274.

Kim, M., Song, N. R., Hong, J., Lee, J.& Pyo, H. (2013). Quantitative analysis of organochlorine pesticides in human serum using headspace solid-phase microextraction coupled with gas chromatography–mass spectrometry. *Chemosphere*, *92*, 279-285.

Kirchner, M., Matisová, E., Hrouzková, S. & de Zeeuw, J. (2005). Possibilities and limitations of quadrupole mass spectrometric detector in fast gas chromatography. *J. Chromatogr. A*, *1090*, 126-132.

Kjeldsen, L. S., Ghisari, M. & Bonefeld-Jorgensen, E. C. (2013). Currently used pesticides and their mixtures affect the function of sex hormone receptors and aromatase enzyme activity. *Toxicol. Appl. Pharmacol.*, *272*, 453-464.

Knez, Ž., Škerget, M.& KnezHrnčič, M. (2013). Principles of supercritical fluid extraction and applications in the food, beverage and nutraceutical industries. Separation, Extraction and Concentration Processes in the Food, Beverage and Nutraceutical Industries, Edited by: Rizvi, S., ISBN: 978-1-84569-645-02013, chpt. 1, pp. 3-38, Woodhead Publishing Limited.

Korba, K., Pelit, L., Pelit, F. O., Ozdokur, K. V., Ertas, H., Eroglub, A. E. & Ertas, F. N. (2013). Preparation and characterization of sodium dodecyl sulfate doped polypyrrole solid phase micro extraction fiber and its application to endocrine disruptor pesticide analysis. *J. Chromatogr. B*, *929*, 90-96.

Kroon, F. J., Hook, S., Jones, D., Metcalfe, S.& Osborn, H. L. (2014). Effects of atrazine on endocrinology and physiology in Juvenile Barramundi, Lates Calcarifer (bloch). *Env. Tox. Chem.*, *33*, 1607-1614.

Kruve, A., Leito, I. & Herodes, K. (2009). Combating matrix effects in LC/ESI/MS: The extrapolative dilution approach. *Anal. Chim. Acta*, *651*, 75-80.

Laborie, S., Moreau-Guigon, E., Alliot, F., Desportes, A., Oziol, L.& Chevreuil, M. (2016). A new analytical protocol for the determination of 62 endocrine-disrupting compounds in indoor air. *Talanta*, *147*, 132-141.

LaFleur, A. D. & Schug, K. A. (2011). A review of separation methods for the determination of estrogens and plastics-derived estrogen mimics from aqueous systems. *Anal. Chim. Acta*, *696*, 6-26.

Lang, Q., Wai, C. M., (2001). Supercritical fluid extraction in herbal and natural product studies—a practical review. *Talanta*, *53*, 771-782.

Lebovka, N., Vorobiev, E. & Chemat, F. (2012). Enhancing Extraction Processes in the Food Industry. CRC Press, Boca Raton, FL.

Lehotay, S. J., Son, K. A., Kwon, H., Koesukwiwat, U., Fu, W., Maštovská, K., Hoh, E.& Leepipatpiboon, N. (2010). Comparison of QuEChERS sample preparation methods for the analysis of pesticide residues in fruits and vegetables. *J. Chromatogr. A*, *1217*, 2548-2560.

Lestremau, F., Willemin, M. E., Chatellier, C., Desmots, S.& Brochot, C. (2014). Determination of cis-permethrin, trans-permethrin and associated metabolites in rat blood and organs by gas chromatography–ion trap mass spectrometry. *Anal. Bioanal. Chem.*, *406*, 3477-3487.

Li, C. M., Jin, F., Yu, Z. Y., Qi, Y. M., Shi, X. M., Wang, M., Shao, H., Jin, M. J., Wang, J. & Yang, M. Q. (2012). Rapid determination of chlormequat in meat by dispersive solid phase extraction and hydrophilic interaction liquid chromatography (HILIC)-electrospray tandem mass spectrometry. *J. Agricult. Food Chem.*, *60*, 6816-6822.

Li, P., Yang, X., Miao, H., Zhao, Y., Liu, W.& Wu, Y. (2013). Simultaneous determination of 19 triazine pesticides and degradation products in processed cereal samples from Chinese total diet study by isotope dilution–high performance liquid chromatography–linear ion trap mass spectrometry. *Anal. Chim. Acta*, *781*, 63-71.

Li, J., Liu, D., Wu, T., Zhao, W., Zhou, Z.& Wang, P. (2014). A simplified procedure for the determination of organochlorine pesticides and polychlorobiphenyls in edible vegetable oils. *Anal. Methods*, *151*, 47-52.

Locatelli, M., Sciascia, F., Cifelli, R., Malatesta, L., Bruni, P. & Croce, F. (2016). Analytical methods for the endocrine disruptor compounds determination in environmental water samples. *J. Chromatogr. A*, *1434*, 1-18.

Lozano, A., Rajski, Ł., Uclés, S., Belmonte-Valles, N., Mezcua, M.& Fernández-Alba, A. R. (2014). Evaluation of zirconium dioxide-based sorbents to decrease the matrix effect in avocado and almond multiresidue pesticide analysis followed by gas chromatography tandem mass spectrometry. *Talanta, 118*, 68-83.

Lu, M., Du, J., Zhou, P., Chen, H., Lu, C.& Zhang, Q. (2015). Endocrine disrupting potential of fipronil and its metabolite in reporter gene assays. *Chemosphere, 120*, 246-251.

Luo, Y. B., Yuan, B. F., Yu, Q. W.& Feng, Y. Q. (2012). Substrateless graphene fiber: a sorbent for solid-phase microextraction. *J. Chromatogr. A, 1268*, 9-15.

Maggioni, S., Balaguer, P., Chiozzotto, C. & Benfenati, E. (2013). Screening of endocrine-disrupting phenols, herbicides, steroid estrogens, and estrogenicity in drinking water from the waterworks of 35 Italian cities and from PET-bottled mineral water. *Environ. Sci. Pollut. Res., 20*, 1649-1660.

Martínez, C., Ramírez, N., Gómez, V., Pocurull, E.& Borrull, F. (2013). Simultaneous determination of 76 micropollutants in water samples by headspace solid phase microextraction and gas chromatography–mass spectrometry. *Talanta, 116*, 937-945.

Mansilha, C., Melo, A., Rebelo, H., Ferreira, I. M., Pinho, O., Domingues, V., Pinho, C.& Gameiro, P. (2010). Quantification of endocrine disruptors and pesticides in water by gas chromatography-tandem mass spectrometry: Method validation using weighted linear regression schemes. *J. Chromatogr. A, 1217*, 6681-6691.

Matisová, E.& Hrouzková, S. (2012). Analysis of Endocrine Disrupting Pesticides by Capillary GC with Mass Spectrometric Detection.*Int. J. Environ. Res. & Publ. Health, 9*, 3166-3196.

Mavedati, O., Bandariyan, E., Aminashayeri, S., Sergez, T. M.& Beigi, B. A. (2015). β-Cryptoxanthin ameliorates the reproductive toxicity of chlorpyrifos in male rabbit. *Comp. Clin. Pathol., 24*, 409-415.

Mnif, W., Hassine, A. I. H, Bouaziz, A., Bartegi, A., Thomas, O.& Roig, B. (2011). Effect of Endocrine Disruptor Pesticides: A Review. *Internat. J. Environ. Res. Publ. Health, 8*, 2265-2303.

Moein, M. M., Abdel-Rehim, A.& Abdel-Rehim, M. (2015). Microextraction by packed sorbent (MEPS). *TrAC Trends in Anal. Chem., 67*, 34-44.

Munaretto, J. S., Ferronato, G., Ribeiro, L. C., Martins, M. L., Adaime, M. B.& Zanella, R. (2013). Development of amultiresidue method for the determination of endocrine disrupters in fish fillet using gaschromatography–triple quadrupole tandem mass spectrometry. *Talanta, 116*, 827-834.

Odendaal, C., Seaman, M. T., Kemp, G., Patterton, H. E.& Patterton, H. G. (2015). An LC-MS/MS based survey of contaminants of emerging concern in drinking water in South Africa. *S. Afr. J. Sci.*, *111*, 160-165.

Oguz, A. R.& Kankaya, E. (2013). Determination of Selected Endocrine Disrupting Chemicals in Lake Van, Turkey. *Bull. Environ. Contam. Toxicol.*, *91*, 283-286.

Pakade, Y. B. & Tewary, D. K. (2010). Development and applications of single-drop microextraction for pesticide residue analysis: a review. *J. Sep. Sci.*, *33*, 3683-3691.

Páleníková, A.& Hrouzková, S. (2014). Microextraction in Packed Syringe: Solvent–Minimized Sample Preparation Technique. *Monatsh. Chem. – Chem. Month.*, *145*, 537-549.

Páleníková, A., Martínez-Domínguez, G., Arrebola, F. J., Romero-González, R., Hrouzková, S. & Frenich, G. A. (2015)[a]. Multifamily determination of pesticide residues in soya-based nutraceutical products by GC/MS-MS. *Food Chem.*, *173*, 796-807.

Páleníková, A., Martínez-Domínguez, G., Arrebola F. J., Romero-González R., Hrouzková, S. & Garrido Frenich, A. (2015)[b]. Occurrence of pesticide residues and transformation products in different types of dietary supplements. *Food Addit.&Contam.: Part A*, *32*, 849-856.

Pedersen-Bjergarrd, S.& Rasmussen, K. E. (1999). Liquid−Liquid−Liquid microextraction for sample preparation of biological fluids prior to capillary electrophoresis. *Anal. Chem.*, *71*, 2650-2656.

Pelit, F. O., Ertas, H., Seyrani, I. & Ertas, F. N. (2013). Assessment of DFG-S19 method for the determination of common endocrine disruptor pesticides in wine samples with an estimation of the uncertainty of the analytical results. *Food Chem.*, *138*, 54-61.

Pelit, L.& Dizdas, T. N. (2013). Preparation and application of a polythiophene solid-phase microextraction fiber for the determination of endocrine-disruptor pesticides in well waters. *J. Sep. Sci.*, *36*, 3234-3241.

Pelit, F. O., Pelit, L., Alaca, C., Ertas, H.& Ertas, F. N. (2014).Preconcentration and determination of endocrine disruptor pesticides in well water by solidified floating organic drop microextraction. *Clean- Soil Air Water*, *42*, 1284-1291.

Pelit, F. O.& Yengin, C.(2014). Application of solidified floating organic drop microextractionmethod for biomonitoring of chlorpyrifos and its oxon metabolitein urine samples. *J. Chromatogr. B*, *949*-950, 109-114.

Piazza, Y., Pandolfi, M., Cuña, R. D., Genovese, G.& Nostro, F. L. (2015). Endosulfan affects GnRH cells in sexually differentiated juveniles of the perciform Cichlasoma dimerus. *Ecotox. Environ. Safe.*, *116*, 150-159.

Pintado-Herrera, M. G., González-Mazo, E.& Lara-Martín, P. A. (2014). Atmospheric pressure gas chromatography–time-of-flight-mass spectrometry (APGC–ToF-MS) for the determination of regulated and emerging contaminants in aqueous samples after stir bar sorptive extraction (SBSE). *Anal. Chim. Acta*, *851*, 1-13.

Pintado-Herrera, M. G., González-Mazo, E.& Lara-Martín, P. A. (2016). In-cell clean-up pressurized liquid extraction and gaschromatography–tandem mass spectrometry determination ofhydrophobic persistent and emerging organic pollutants in coastalsediments. *J. Chromatogr. A*, *1429*, 107-118.

Prieto, A., Basauri, O., Rodil, R., Usobiaga, A., Fernandez, L. A., Etxebarria, N.& Zuloaga, O. (2010). Stir-bar sorptive extraction: A view on method optimisation, novel applications, limitations and potential solutions. *J. Chromatogr. A*, *1217*, 2642-2666.

Purdešová, A., Hrouzková, S., Andraščíková, M., Krascsenits, Z.& Matisová, E. (2013). Evaluation of calibration approaches for quantification of pesticide residues in surface water by SPE with small-size cartridges followed by fast GC-MS. *Anal. Methods*, *5*, 3403-3409.

Qin, Y., Zhao, P., Fan, S., Han, Y., Li, Y., Zou, N., Song, S., Zhang, Y., Li, F., Li, X.& Pan, C. (2015). The comparison of dispersive solid phase extraction and multi-plug filtration cleanup method based on multi-walled carbon nanotubes for pesticides multi-residue analysis by liquid chromatography tandem mass spectrometry. *J. Chromatogr. A*, *1385*, 1-11.

Ramos, L. (2012). Basics and Advances in Sampling and Sample Preparations. In Chemical Analysis of Food: Techniques and Applications, Eds. Y. Picó, Elsevier, Waltham MA, USA, pp. 59-90.

Rathore, H. S.& Nollet, L. M. L. (2012). Pesticides evaluation of environmental pollution. 1st Edition, Taylor & Francis Group, Boca Raton, USA.

REACH 1907, (2006). Regulation (EC) No. 1907/2006 of the European Parliament and of the Council of 18 December 2006 concerning the Registration, Evaluation, Authorisation and Restriction of Chemicals (REACH), establishing a European Chemicals Agency. Available on-line http://www.hse.gov.uk/reach/reachtext.pdf. Accessed 8 March 2016.

Rezaee, M., Yamini, Y.& Faraji, M. (2010). Evolution of dispersive liquid–liquid microextraction method. *J. Chromatogr. A* 1217, 2342-2357.

Rogers, J. A., Metz, L. & Yong, V. W. (2013). Review: Endocrine disrupting chemicals and immune responses: A focus on bisphenol-A and its potential mechanisms. *Mol. Immunol.*, *53*, 421-430.

Ros, O., Izaguirre, J. K., Olivares, M., Bizarro, C., Ortiz-Zarragoitia, M., Cajaraville, M. P., Etxebarria, N., Prieto, A.& Vallejo, A. (2015). Determination of endocrine disrupting compounds and their metabolites in fish bile. *Sci. Total Environ.*, *536*, 261-267.

Salvatierra-Stamp, V. C., Ceballos-Magaña, S. G., Gonzalez, J., Ibarra-Galván, V.& Muñiz-Valencia, R. (2015). Analytical method development for the determination of emerging contaminants in water using supercritical-fluid chromatography coupled with diode-array detection. *Anal. Bioanal. Chem.*, *407*, 4219-4226.

Sanchez-Prado, L., Garcia-Jares, C., Dagnac, T.& Llompart, M. (2015). Microwave-assisted extraction of emerging pollutants in environmental and biological samples before chromatographic determination, *Trends in Anal. Chem.*, *71*, 119-143.

SANTE document, (2015). Guidance document on analytical quality control and method validation procedures for pe residues analysis in food and feed. SANTE/11945/2015, European Commission, Directorate-general for Health and Food Safety, Safety of the Food Chain Chemicals, Pesticides and Biocides.

Santhi, V. A., Hairin, T.& Mustafa, A. M. (2012). Simultaneous determination of organochlorine pesticides and bisphenol A in edible marine biota by GC–MS. *Chemosphere*, *86*, 1066-1071.

Santos, H. M.& Capelo, J. L.(2007). Trends in ultrasonic-based equipment for analytical sample treatment. *Talanta*, *73*, 795-802.

Sarafraz-Yazdi, A.& Amiri, A. (2010). Liquid-phase microextraction. *Trends in Anal. Chem.*, *29*, 1-14.

Satpathy, G., Tyagi, Y. K.& Gupta, R. K. (2011). A Novel Optimised and Validated Method for Analyiss of Multi-Residues of Pesticides in Fruits and Vegetables by Microwave-Assisted Extraction (MAE)-Dispersive Solid Phase Extraction (dSPE)-Retention Time Locked (RTL)-Gas Chromatography-Mass Spectrometry with Deconvolution Reporting Software (DRS). *Food Chem.*, *127*, 1300-1308.

Shaw, I. (2009). Endocrine-disrupting chemicals. Woodhead Publishing Lim and CRC Press, Boca Raton.

Shimelis, O., Yang, Y., Stenerson, K., Kaneko, T.& Ye, M. (2007). Evaluation of a solid-phase extraction dual-layer carbon/primary secondary amine for clean-up of fatty acid matrix components from food extracts in multiresidue pesticide analysis. *J. Chromatogr. A*, *1165*, 18-25.

Sichilongo, K. & Banda, D. (2013). GC-MS determination of targeted pesticides in environmental samples from the Kafue flats of Zambia. *Bull. Environ. Contam. Toxicol.*, *91*, 510-516.

Smyth, W. F. (2003). Electrospray ionisation mass spectrometric behaviour of selected drugs and their metabolites. *Anal. Chim. Acta*, *492*, 1-16.

Sosa-Ferrera, Z., Mahugo-Santana, C.& Juan Santana-Rodríguez, J. (2013). Analytical Methodologies for the Determination of Endocrine Disrupting Compounds in Biological and Environmental Samples. *BioMed. Res. Internat.*, 2013, 1-23.

Souza-Silva, É. A., Gionfriddo, E.& Pawliszyn, J. (2015). A critical review of the state of the art of solid-phase microextraction of complex matrices II. Food analysis. *Trends in Anal. Chem.*, *71*, 236-248.

Subedi, B., Aguilar, B., Robinson, E. M., Hageman, K. J., Björklund, E., Sheesley, R. J.& Usenko, S.(2015). Selective pressurized liquid extraction as a sample-preparation technique for persistent organic pollutants and contaminants of emerging concern. *Trends in Anal. Chem.*, *68*, 119-132.

Sun, H., Ge, X., Iv, Y. & Wang, A. (2012). Application of accelerated solvent extraction in the analysis of organic contaminants, bioactive and nutritional compounds in food and feed. *J. Chromatogr. A*, *1237*, 1-23.

Swedenborg, E., Rüegg, J. I., Makela, S.& Pongratz, I. (2009). Endocrine disruptive chemicals: mechanisms of action and involvement in metabolic disorders. *J. Molecular Endocrinol.*, *43*, 1-10.

Synaridou, M. E. S., Sakkas, V. A., Stalikas, C. D.& Albanis, T. A. (2014). Evaluation of magnetic nanoparticles to serve as solid-phase extraction sorbents for the determination of endocrine disruptors in milk samples by gas chromatography mass spectrometry. *J. Chromatogr. A*, *1348*, 71-79.

Tournier, M., Pouech, C., Quignot, N., Lafay, F., Wiest, L., Lemazurier, E., Cren-Olivé, C. & Vulliet, E. (2015). Determination of endocrine disruptors and endogenic androgens and estrogens in rat serum by high-performance liquid chromatography-tandem mass spectrometry. *Steroids*, *104*, 252-262.

US EPA, Document available on-line: http://www.epa.gov/endocrine-disruption/endocrine-disruptor-screening-program-edsp-overview, (accessed on 20 february 2016).

Uygun, Z. O. & Dilgin, Y. (2013). A novel impedimetric sensor based on molecularly imprintedpolypyrrole modified pencil graphite electrode for trace leveldetermination of chlorpyrifos. *Sens. Act. B: Chem.*, *188*, 78-84.

Vaithinathan, S., Saradha, B.& Mathur, P. P. (2010). Methoxychlor induces apoptosis via mitochondria- and FasL-mediated pathways in adult rat testis. *Chem. Biol. Interact.*, *185*, 110-118.

Valverde, A., Aguilera, A., Rodriguez, M.& Brotons, M. (2009). Evaluation of a multiresidue method for pesticides in cereals using supercritical fluid extraction and gas chromatographic detection. *J. Environ. Sci. Health, Part B*, *44*, 204-213.

Vandenberg, L. N., Colborn, T., Hayes, T. B., Heindel, J. J., Jacobs, D. R., Jr., Lee D. H., Shioda, T., Soto, A. M., vom Saal F. S., Welshons W. V. R., Zoeller, T.& Peterson Myers, J. (2012). Hormones and Endocrine-Disrupting Chemicals: Low-Dose Effects and Nonmonotonic Dose Responses. *Endocr Rev.*, *33*, 378-455.

Vazquez-Roig, P. & Picó, Y. (2015). Pressurized liquid extraction of organic contaminants in environmental and food samples. *Trends in Anal. Chem.*, *71*, 55-64.

Wang, H., Ding, J.& Ren, N. (2016). Recent advances in microwave-assisted extraction of trace organic pollutants from food and environmental samples. *Trends in Anal. Chem.*, *75*, 197-208.

WHO document. (2002). Available on-line: http://www.who.int/ipcs/publications/en/ch1.pdf, Accessed 15 February 2016.

Wilkowska, A. & Biziuk, M. (2011). Determination of pesticide residues in food matrices using the QuEChERS methodology. *Food Chem.*, *125*, 803-812.

Yang, E. Y. & Shin, H. S. (2013). Trace level determinations of carbamate pesticides in surface water by gas chromatography–mass spectrometry after derivatization with 9-xanthydrol. *J. Chromatogr. A*, *1305*, 328-332.

Yang, Y., Kong, W., Zhao, L., Xia, Q., Liu, H., Zhao, X.& Yang, M. (2015). A multiresidue method for simultaneous determination of 44 organophosphorous pesticides in Pogostemon cablin and relatedproducts using modified QuEChERS sample preparation procedureand GC–FPD. *J. Chromatogr. B*, *974*, 118-125.

Zhang, X., Tian, H., Wang, W.& Ru, S. (2013). Exposure to monocrotophos pesticide causes disruption of the hypothalamic–pituitary–thyroid axis in adult male goldfish (Carassius auratus). *Gen. Comp. Endocr.*, *193*, 158-166.

Zhang, Y.& Lee, H. K. (2012). Application of ultrasound-assisted emulsification microextraction based on applying low-density organic solvent for the determination of organochlorine pesticides in water samples. *J. Chromatogr. A*, *1252*, 67-73.

Zhao, E., Shan, W., Jiang, S., Liu, Y.& Zhou, Z. (2006). Determination of the chloroacetanilide herbicides in waters using single-drop microextraction and gas chromatography. *Microchem. J.*, *83*, 105-110.

BIOGRAPHICAL SKETCH

Svetlana Hrouzková is Associate Professor of Analytical Chemistry at Slovak University of Technology in Bratislava, Faculty of Chemical Technology and Food. Her research is focused on developing new analytical methods for determining pesticide residues in food and environmental samples, using chromatographic techniques and mass spectrometry detection, with emphasis on the treatment of sample systems, especially on microextraction techniques. She is the author of 35 articles in international peer reviewed journals, 6 chapters in monographs, 113 symposium papers and co-authored 3 university textbooks.

Agneša Páleníková is graduating PhD. student in Institute of Analytical Chemistry at Slovak University of Technology in Bratislava, Faculty of Chemical and Food Technology. At present she works on method development in pesticide residues analysis in food and pharmaceutical samples, with emphasis on sample preparation techniques. She is the author of 5 articles in international journals, 1 chapter in monographs and 33 conference papers.

Reviewed by Prof. Eva Matisová, DrSc., professor emeritus at Slovak University of Technology in Bratislava, Faculty of Chemical and Food Technology, Slovakia.

Dr. Eva Korenkova, Mass Spectrometry Specialist, Government of Ontario, Canada.

In: Endocrine Disrupting Chemicals
Editor: Sabrina Johnston
ISBN: 978-1-63485-231-9

Chapter 2

TRUMPED BY TRADITIONAL HAZARDS: EDC EXPOSURE FROM RECYCLED WASTEWATER COMPARED TO OTHER WATER RELATED RISKS IN DEVELOPED AND DEVELOPING NATIONS

Vivian Futran Fuhrman[1,*], Alon Tal[1], Alfred Abed-Rabbo[2], Nader el-Khateeb[2], Nina Gordon-Kirsch[1], Tal Godinger[3], Wa`d Odeh[3] and Shai Arnon[3]

[1]Institute for Dryland, Environmental and Desert Research,
The Jacob Blaustein Institutes for Desert Research,
Ben-Gurion University of the Negev, Sede Boqer, Israel
[2]Water and Environmental Development Organization (WEDO),
Bethlehem University, Bethlehem, Israel
[3]Zuckerberg Institute for Water Research,
The Jacob Blaustein Institutes for Desert Research,
Ben-Gurion University of the Negev, Sede Boqer, Israel

* Corresponding Author: Vivian.Futran@gmail.com.

ABSTRACT

Israel is the global leader in wastewater reuse making it the ideal test case to investigate human exposure to contaminants via recycled wastewater pathways. Studies around the world have revealed that wastewater frequently contains endocrine disrupting chemicals (EDCs) that are not completely removed by traditional treatment. As these compounds can harm ecological and human health, it was initially hypothesized that residual EDCs were contributing to reproductive problems in Israel's population. The fate of EDCs from wastewater effluent in Israel and the Palestinian West Bank's shared water basins was assessed over a two year period, and the associated health risks and economic costs of competing wastewater treatment options evaluated. Results from this analysis can inform both developed and developing nations since Israel highly treats and reuses most of its wastewater while only a small percent of sewage undergoes any treatment in the Palestinian territories. EDCs in the region's aquatic environment were measured at concentrations consistent with global averages. According to modeling results, human exposure to bisphenol A via recycled wastewater is likely small compared to ingestion through food. While a potential health risk was identified from bisphenol A in recycled wastewater, it was most pronounced in extreme exposure situations and when reproductive capacity was already compromised. In the West Bank, a cost-effectiveness analysis determined that the marginal improvement in removal rates of estrogens from building tertiary versus secondary level wastewater treatment facilities would not justify the added cost to the Palestinian Authority. Secondary treatment of sewage removed EDCs sufficiently, and the money saved by not investing in tertiary treatment could be better used to provide more essential sanitation services. Tertiary treatment produces other important environmental benefits in which countries like Israel may wish to invest, but it yields only marginally better EDC removal, so developed countries aiming to reduce overall EDC exposure should target sources other than treated wastewater. These findings suggest that EDCs from treated wastewater merit continued research, and demonstrate the importance of environmental cooperation in regions with bifurcated development. However, as EDCs are not the biggest risk to human health in developed or developing nations at this time, they should not be the primary driver for policy governing wastewater management.

INTRODUCTION

As water scarcity threatens a growing number of regions worldwide, more countries are turning to reuse of treated wastewater effluent to balance their water budget even while the environmental and health consequences of unregulated organic contaminants in this source of water are still under investigation. For example, in Israel ambitious government goals have driven innovation in water management and treatment technology [1] helping position it as a leader in the field [2]. Israel recycles its municipal wastewater at a globally unprecedented rate, recapturing 96% for treatment to at least secondary levels and reusing 86% [1]. This is more than 80 times the rate of <1% wastewater recycled by the US [1, 3, 4], and over four times the second highest rate of about 20% in Spain [1]. While an "integrated water management" system has helped Israel overcome an arid climate and a limited supply of fresh water, pioneering technologies like wastewater reuse have raised concerns regarding unknown health implications.

At the same time, neighboring communities in the Palestinian territories face exposure to contaminants in untreated and poorly treated wastewater. The Palestinian Authority (PA) still faces chronic shortages in water availability and acute water quality risks including exposure to microbial pathogens and other traditional contaminants [5, 6], even as emerging contaminants such as endocrine disrupting chemicals (EDCs) gain increasing attention on the international stage.

EDCs are a heterogeneous category of "chemicals of emerging concern" commonly found in wastewater systems [7, 8]. While hormones are endogenously produced and emitted by plants, animals, and humans, natural concentrations in the environment are augmented by anthropogenic sources of synthetic EDCs found in pharmaceuticals, personal care products, foods, plasticizers, pesticides, and other household, industrial, and agricultural products [9–12]. These can leak directly into the aquatic environment or be discharged into the sewage system by humans where their fate depends on the type of treatment the wastewater receives [13]. Today EDCs are ubiquitous in aquatic environments [10, 14].

EDCs mimic or disrupt hormone function regulated by the endocrine system [15]. Exposure to EDCs can impair normal endocrine system functioning via estrogen, androgen, or thyroid pathways, disrupting various related networks and processes in the body [15]. This can damage wildlife [16–19] and has the potential to damage human health [20–24]. EDCs have been linked to a range of problems including adverse reproductive and developmental health effects in males and females, congenital malformations, hormone mediated cancers, and metabolic processes among others [25]. Concentrations of EDCs as low as parts-per-trillion can disturb endocrine processes [26–28], especially for vulnerable subpopulations in hormone-driven developmental stages such as pregnancy, adolescence, and menopause [29–32].

Health risks associated with EDCs will become easier to assess as the vast number of chemicals with endocrine potential are evaluated and better characterized. In the meantime, scientists and risk assessors can still advise the policy community regarding best practices for avoiding dangerous exposure to EDCs. The goal of this analysis is to investigate the transport and fate of EDCs from wastewater effluent in Israel and the West Bank to determine whether they pose a risk to human health. It is especially germane in this region as Israel's effluent reuse rate is unparalleled while Palestinians still discharge copious amounts of sewage, often untreated, into the ambient environment. In evaluating the resulting hydrological, economic, and health consequences, this study seeks to provide a clearer picture regarding whether EDCs should be a driver for public policy in both developed and developing nations. It also highlights important management strategies to protect environments shared by countries in close proximity with extremely different economic realities.

The chapter presented here was completed as part of a PhD dissertation under a larger joint research initiative between Ben-Gurion University of the Negev and The Water and Environmental Development Organization (WEDO), and supported by the USAID Middle East Regional Cooperation (MERC) Program (grant TA-MOU-11-M31-015). The primary objective of this MERC project was the collaboration of a joint Israeli-Palestinian research team in identifying and measuring endocrine disrupting compounds (EDCs) in the aquatic environment of Israel and the Palestinian West Bank at different stages in the wastewater treatment process, as well as in different environments that have a long history of receiving wastewater discharges in the region (e.g., streams and irrigation). The goal of the project was to take the field (water sampling) and laboratory (chemical analysis) results, and translate these raw data into actionable recommendations for the region. This called for a combination of methodologies from various disciplines: public health/risk assessment, economics, and policy assessment.

BACKGROUND

Wastewater Magangement in Israel and the West Bank

In Israel, almost all municipal wastewater is recaptured, directed to wastewater treatment facilities, and treated to a minimum of secondary (biological) treatment levels, or even to tertiary (chemical and advanced) treatment levels. Israel currently reuses most of the 500 million cubic meters per year of municipal wastewater it produces [33]. Effluents are then allocated for industrial uses, stream rehabilitation, landscaping, and above all agricultural irrigation. Already 50% of Israeli agricultural irrigation water came from recycled sewage in 2014, and Israel plans to increase its rate of municipal wastewater reuse from 86% to 90% over the next 5-7 years [1]. To keep up with increasing consumption and met these goals, Israel plans to augment artificial water sources (desalination, seawater and imports), doubling net allocations to agricultural irrigation by 2050 [34] without "risk to soil and water sources" [35].

Although they share a border, as well as most of their surface catchments and major groundwater aquifers, Israel and the Palestinian territories face extremely different realities with regards to water resources. While Israel is a developed country with highly advanced national water management and governance systems [54], the PA—in charge of most of the populated areas in the West Bank—is considered an economically developing country. Palestinians face very limited water resources and physical infrastructure, weak institutions and operators, political hurdles, stressed finances, and demographic growth [55, 56]. The only wastewater reuse that occurs is local, small scale, and considered a negligible amount [56]. Consequently, there is a deficit of both water and sanitation infrastructure, services, and operations in the PA [57]. Israel captures and treats approximately 15-17 million cubic meters of the PA's sewage water every year at the expense of the PA [6, 58]. This water is then sold and reused inside of Israel without payment to the PA for the value of the natural resource [6].

As of 2012, only 31% of the Palestinian population was connected to a wastewater collection network [56]. The water infrastructure suffered an average of 30% loss from leakage [56] and included only two fully functioning municipal wastewater treatment plants (WWTPs) in 2014 [59]. Subsequently, 65% of Palestinian sewage is collected in cesspools that leach into the ground while 33% is dumped directly into dry river beds that flow freely throughout the year [56]. This untreated wastewater can percolate into the ground, contaminating aquifers [5, 56, 60]. At present, there are streams in the West Bank comprised solely of wastewater [61] which release foul smelling odors [62] and can expose locals to pathogens from human and industrial waste [5]. Many of these streams flow into Israel as the West Bank is topographically located at the head of the catchment, which contributes to the long lists of grievances the Palestinian and Israeli governments have against each other with regards to shared water resources.

Reproductive Health and Evidence of Emerging Contaminants

Studies around the world have demonstrated that EDCs are commonly found in wastewater [36–39]. Many of these compounds are not completely removed by primary, secondary, or sometimes even tertiary wastewater treatment [40–43] and can only be eliminated through advanced treatment techniques [30, 44]. It was therefore hypothesized that wastewater treated to conventional levels in Israel might be contributing to the release of trace amounts of EDCs in effluent [10, 45–47]. Once in the environment, these compounds can make their way into fresh water where they are increasingly ubiquitous [10, 48–50], or if used for irrigation, can be taken up by crop plants [51–53]. This recognition elicited alarm as to whether the concentration of EDCs in Israel's aquatic environment might be increasing with every new cycle of use, increasing human exposure and threatening health.

Growing reproductive health problems in Israel suggested there may be meaningful environmental exposures to EDCs in the region. Evidence includes a decreased age of menarche in female army recruits [63], a steady decline in the age of puberty among the general population of Israeli girls [64], an increase in the prevalence of male factor infertility [65–67], an increase in testicular cancer rates [68], a reduction in male sperm count and motility [69], a decrease in normal sperm morphology among sperm donors [70], and a decrease in sperm motility and concentration among sperm donors [71]. If these trends continue, the average Israeli male could be characterized as "reproductively impaired" by 2020 according to World Health Organization (WHO) criteria [72]. Despite possible confounders and other limitations associated with this data, it was initially suspected that emerging contaminants in treated wastewater effluent such as EDCs could be partially responsible for deleterious health trends [72, 73].

Previous studies in Israel have detected contaminants of emerging concern across various environmental media. In the 1980's and 1990's, high concentrations of industrial solvents were detected from wells beneath a field that had been irrigated with recycled wastewater [74–76]. In 2008, Chefetz et al., reported pharmaceutical compounds in secondary treated wastewater at concentrations similar to those described in wastewater effluents, and described their movement through irrigated soil. Avisar et al., studied the long term impact of wastewater irrigation and found that high concentrations of antibiotics had moved through soil and into the water table below [77], which added to a list of environmental antibiotics and metabolites identified locally [78]. Using carbamazepine as tracer, Gasser et al., [79] demonstrated percolation into a section of the Israeli Coastal Aquifer from a wastewater effluent recharge system nearby. In 2011, Shenker et al., found that pharmaceuticals could be taken up by crops irrigated with reclaimed wastewater at realistic levels. Finally, in 2014 Malchi et al., discovered measurable levels of pharmaceutical compounds in root vegetables including varieties eaten raw. For example, there were 5-10 nanograms per grams of carbamazepine found in carrots (with the precise amount depending on the soil treatment). Although these studies did not focus on EDCs, they demonstrated the leaching potential and subsequent risk of organic compounds originating in wastewater.

Previous Regional Investigation of EDCs

Despite the steady stream of research evaluating the environmental effects of wastewater reuse in Israel, no similar studies of wastewater's effect on Palestinians had been undertaken heretofore. This study was undertaken to fill that gap, as well as to provide regional information on EDCs in the aquatic environment. Preceding this study, there had been minimal data collected on EDCs in this region [80–82] even though these compounds are of growing global concern. A comprehensive survey of EDC concentrations in the region's water resources had yet to be conducted, and there had not been an assessment undertaken of these chemicals' presence following different treatment technologies. In addition, there had not been a transboundary project systematically assessing the fate and implications of EDCs from wastewater throughout the Palestinian/Israeli shared aquatic environment.

The physical proximity of Israel and the PA makes it possible to study whether EDCs should be a water policy priority in developed nations, developing nations, and areas straddling both conditions. Such an asymmetrical geographic juxtaposition is common throughout the world, with extreme economic and environmental disparities found between contiguous countries such as the US and Mexico, the Dominican Republic and Haiti, North and South Korea, etc., as well as areas like Eastern and Western Europe. As such, this study has far-reaching relevance.

ANALYSES

Hydrological Sampling and Chemical Analysis

In 2012, the study presented here was launched to investigate the fate and transport of EDCs from recycled wastewater in the aquatic environment of Israel and the West Bank. The project was a collaborative effort between Israeli and Palestinian researchers utilizing a range of expertise, and the first such project focusing on EDCs. The study was designed to be the most comprehensive investigation of EDCs in the region to date, and the first in the context of recycled wastewater. Four water sampling campaigns were conducted in 2013 and 2014 (two summer and two winter) covering 60 locations across Israel and the West Bank. The samples reflected various treatment processes from raw sewage to primary, secondary, and tertiary treatments, all the way through discharge into the irrigation water system or into local streams and groundwater. All water samples underwent chemical analysis at the laboratories of Israel's Ministry of Health using U.S. Environmental Protection Agency (EPA) standard methods [83].

As no significant seasonal differences were detected in between measurements, the concentration values were averaged across the three campaigns included in analysis (summer 2013, winter 2014, and summer 2014). In total, five EDCs were detected in Israel including endogenous estrogenic hormones (estriol, estrone), xenoestrogens (bisphenol A and 4-tert-octylphenol), and one estrogenic compound used in personal care products (triclosan) (See Table 1 below). One additional estrogenic hormone was detected, 17-beta estradiol, but only below the limit of detection and was thus not included in analysis.

The same estrogenic EDCs were identified in the West Bank. Figure 1 below shows the EDCs detected in influent (raw/untreated) and effluent (treated and released) wastewater at the only two major WWTPs functioning in the West Bank at the time of analysis [59]. Levels of estrone and estriol in raw and treated wastewater were similar across both Palestinian WWTPs, while triclosan was only detected in Al Bireh WWTP, bisphenol A (BPA) was only found in Al Bireh's influent, and 4-tert-octylphenol was not detected in the influent or effluent of either WWTP. Comparing these two WWTPs to the averages across Israeli WWTPs, estrone and estriol in influent were higher in the West Bank—close to or over 100 nanograms per liter (ng/L)— than in Israel where they were under 100 ng/L. Nonetheless, the concentrations in the West Bank were very close to those measured in Israel's effluent (all under 5 ng/L). BPA in the influent of Al Bireh was higher than in Israel by about 500 ng/L, but was reduced to zero in both the West Bank and Israel after treatment. Triclosan levels in the influent and effluent in Al Bireh were comparable to Israeli levels.

Table 1. Range, average, and median concentrations of EDCs measured in raw and treated wastewater in Israeli WWTPs (ng/L)

(Adapted from Godinger [83])

	Raw WW			Treated WW			Limits of Detection
Compound	**Range**	**Average**	**Median**	**Range**	**Average**	**Median**	**Summer '13/Winter '14/Summer '14**
Estriol	0-124	51	46	0-9.4	1.7	0	.03/.05/.05
Estrone	0-181	65	72	0-13	2.1	0.3	.01/.05/.05
BPA*	0-6793	801	238	0	0	0	5/1/1
OP**	0-4960	691	0	0-50	3.3	0	5/1/1
TCS***	162-6800	1272	981	0-400	80	0	5/5/5

*BPA = Bisphenol A.
**OP = 4-tert-octylphenol.
***TCS = Triclosan.

Both Al Bireh and Nablus treat wastewater to secondary levels, whereas five out of the six Israeli WWTPs included in this study treat wastewater to tertiary levels (and the sixth to secondary levels). Nonetheless the secondary technology utilized in the PA appears to be removing EDCs adequately, at rates comparable to those in Israel, despite the significantly higher Israeli investment and capacity.

Of the 60 locations sampled across Israel and the PA, 19 in the PA contained raw wastewater lacking any treatment. The remaining 41 locations consisted of fresh water or wastewater that received some treatment—even if only at a basic level—and were then released. These samples, representing "treated" wastewater effluent as it moved through the environment, were the subject of further examination. (See Appendix A for average concentration of the EDC in these samples). Estradiol equivalencies (EEQ) were calculated for the five EDCs detected in both Israel and the West Bank. EEQs express estrogenic potency relative to estradiol [85–87] which effectively normalized units so that concentrations can be directly compared.

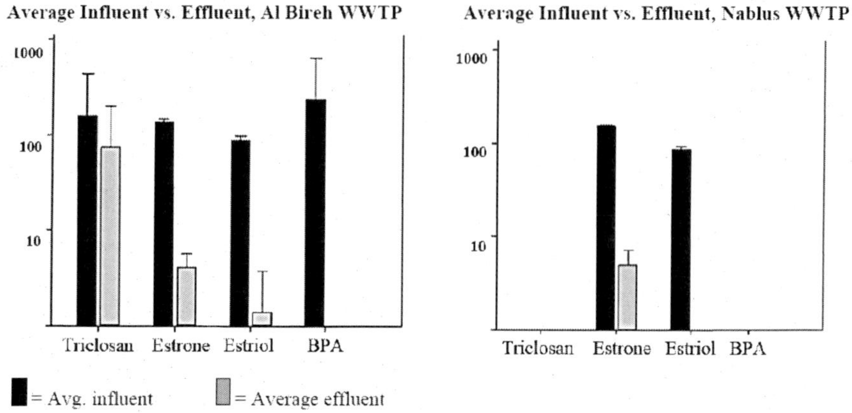

Adapted from Odeh [84].

Figure 1. EDC Concentrations in raw and treated wastewater in PA WWTPs (ng/L).

Estradiol equivalencies (EEQs) are calculated by multiplying each compound concentration by its estradiol equivalency factor (EEF) and is set to one for estradiol [86, 88]. An EEF is defined as half the maximal effective concentration of estradiol divided by half the maximal effective concentration (EC) of another EDC ($EC50_{Estradiol}/EC50_{EDC}$). An EEF is multiplied by a compound's concentration to calculate the compounds estradiol equivalent (EEQ). In this analysis, the EEFs for estrone, estriol, and BPA were borrowed from a paper by Vega-Morales et al., [86] which averaged values from a mix of *in vitro* bioassays to find average EEFs. Triclosan and 4-tert-octylphenol were not included in this paper so the EEF for triclosan was taken from Blavier [89], and for 4-tert-octylphenol an average of EEFs from across the literature [41, 90, 91] was used to approximate the scope of the EEFs used for the other compounds. The EEFs used for each EDC in this analysis are as follows: .11 for estrone and estriol, 5E-08 for triclosan, .00021 for 4-tert-octylphenol, and .00039 for BPA.

It is important to note that EEQs are a toxicological as opposed to a biological concept. The value in finding total EEQ and percent contribution to total EEQ is based on the assumption that the estrogenic potency of these EDCs is always additive which may not be the case. For now, estrogenic bioassays converted into EEQs provide a way to calculate the influence of these compounds, however the data on potency and endpoints of various EDCs is still growing which affects our understanding of EDCs individually and in mixtures [92]. While estrogenic compounds are relatively well studied compared to other EDC groups [93], the assumptions underlying their assessment may change with the advancement of our knowledge in the future.

EEQs were calculated for each compound concentration at the different locations across Israel and the West Bank where treated wastewater effluent was sampled (see Appendix B for average concentration of the EDC expressed as EEQs). The EEQs were averaged across all locations to yield one average EEQ for each compound, and the contribution that each compound made to estrogenicity was expressed as percent of total EEQ. This was done for all treated wastewater samples in the region (Table 2), for the samples from Israel only (Table 3), and for the samples from the West Bank only (Table 4).

Table 2. Estrogenic Equivalencies and Contribution to Total Estrogenicity for Treated Wastewater Samples from Israeli and the West Bank Combined

	Estrone	Estriol	Triclosan	4-tert-Octylphenol	BPA	Total
Average (ng/L)	3.00	2.45	93	3	11	
Standard Deviations	3.90	4.99	149	2	47	
N (number of samples)	41	41	41	27	41	
EEQ (average)	0.3	0.3	4.65E-06	1.28E-03	4.47E-03	0.605
Contribution to Total EEQ	54.56%	44.49%	<.01%	0.21%	0.74%	

The EEQ data facilitate a comparison between the estrogenic contributions of each EDC to help understand their relative impact on health. It is clear from this analysis that estrone and estriol make the largest contribution to estrogenicity across the board. Estrone makes the largest contribution when all locations are considered together (approximately 55%) as well as when Israeli locations are considered in isolation (approximately 61%), and estriol makes the largest contribution in West Bank locations (approximately 69%). This is consistent with findings that endogenous hormones tend to be the main contributors to the estrogenicity of sewage [94–96] as they are much more potent than synthetic EDCs [97]. Estrone and estriol are equally estrogenic relative to estradiol (i.e., both have EFFs of .11), however results here show that treated wastewater in the West Bank contains about twice the concentration of estriol compared to estrone whereas estrone predominates in Israel.

Table 3. Estrogenic Equivalencies and Contribution to Total Estrogenicity for Treated Wastewater Samples from Israel

	Estrone	Estriol	Triclosan	4-tert-Octylphenol	BPA	Total
Average (ng/L)	3.41	2.15	110	3	14	
Standard Deviations	4.21	2.76	162	2	54	
N (number of samples)	32	32	32	18	32	
EEQ (average)	0.4	0.2	5.51E-06	5.33E-04	5.28E-03	0.617
Contribution to Total EEQ	60.76%	38.30%	<.01%	0.09%	0.86%	

Table 4. Estrogenic Equivalencies and Contribution to Total Estrogenicity for Treated Wastewater Samples from the West Bank

	Estrone	Estriol	Triclosan	4-tert-Octylphenol	BPA	Total
Average (ng/L)	1.55	3.50	32	4	4	
Standard Deviations	2.12	9.66	58	1	1	
N (number of samples)	9	9	9	9	9	
EEQ (average)	0.2	0.4	1.59E-06	8.63E-04	1.60E-03	0.558
Contribution to Total EEQ	30.48%	69.08%	<.01%	0.15%	0.29%	

The EEQ data also provide a way to assess the additive estrogenic potency of all five EDCs cumulatively. According to this analysis, after averaging across locations and campaigns and then summing across compounds, the total EEQ was similar for Israel (approximately .62 ng/L EEQs) as for the West Bank (approximately .56 ng/L EEQs). It should be noted that this data may be skewed as there were nearly 4 times more data points for Israel (32 sampling locations) than for the West Bank (9 sampling locations). This may have rendered a more representative picture of Israel than of the West Bank. Future research can improve the resolution of such an evaluation.

Based on the chemical results described above, two additional types of analyses were performed: human health risk and cost-effectiveness.

Exposure Assessment and Relative Risk

For this study on Israel and the West Bank, an assessment of exposure and subsequent health effects was conducted using the chemical analysis results and a specially designed model. EDCs present many challenges to conventional risk assessment methods rendering it imprudent or impossible to calculate *absolute* risk from EDCs. This is in part due to gaps in knowledge regarding EDC functioning, and in part because EDCs often function differently than traditional contaminants [98]. However, risk assessment tools can still be used with caution to evaluate EDCs [99] alongside appropriate risk communication and full transparency regarding steps taken and uncertainty involved. Risk can be estimated relative to other sources or over time, different exposure scenarios weighed, or health outcomes approximated where dose-response curves are available.

The model in this study was designed to calculate exposure to EDCs for a hypothetical scenario, and to predict a potential health impact associated with the given exposure. The exposure and risk associated with BPA— a widely manufactured industrial EDC— were calculated as a "proof of concept" to demonstrate that such a model is effective methodology for calculating exposure, and connecting information to determine a health outcome. The model was designed to be very conservative in order to capture a "worst-case" exposure scenario. It was presumed that if exposure is relatively safe in this scenario, then the general public in a real life scenarios is all the more likely to be safe. As oral ingestion is humans' primary route of exposure to BPA, intake of BPA from recycled wastewater was estimated from a combination of two exposure pathways: drinking groundwater infiltrated by treated wastewater, and eating crops irrigated with treated wastewater. This exposure scenario is hypothetical as none of the well water tested thus far was contaminated. However, the scenario is plausible, especially as water scarcity in the region increases and water demand is not met [58], driving the need to dig water wells even where it is illegal and unregulated. This makes the Palestinian population particularly vulnerable to EDC exposure through this source.

The model demonstrated that, even in a conservative scenario, the exposure to BPA from recycled wastewater through these two sources combined was still slightly less than that calculated by the European Food Safety Authority to come from food in a typical Western diet [100]. This implies that the exposure of the general public to BPA through recycled wastewater is most likely lower than that through diet. To provide context for the calculations of BPA exposure from recycled wastewater, Israeli biomonitoring data of measured concentrations of BPA in the population's urine was used to back-calculate aggregate BPA exposure (from all sources combined). The total BPA exposure was found to be beneath the threshold set forth in the safety standards of the United States and European Union of 4 micrograms per kilogram of body weight per day [100].

Finally, the model was able to connect the exposure estimate with a human health outcome. Using a concentration-response curve established for human males [101], the model calculated the impact had by BPA exposure through treated wastewater on four key semen parameters: concentration, volume, progressive motility, and normal morphology. The apical effect on male fertility was predicted by the model using a modified version of the van der Steeg equation [102] to translate the impact on sperm parameters into effect on chances of fertility within one year. It was found that the health impact on sperm would be appreciable most of all for 1) the most extreme intake situations (i.e., the highest percentiles of exposure), and 2) individuals already demonstrating compromised sperm health (i.e., those individuals in the lowest percentiles of sperm function). (Publication of full model design and findings is forthcoming. Contact authors for more details).

There are limitations inherent in this modeling and assessment, and a case study on BPA cannot guarantee that findings apply to other EDCs. Future epidemiological and biomonitoring studies in the region are necessary to confirm this, and will need to be complimented by improvements in our understanding of EDC characteristics including mechanisms of action, mixture effects, low dose effects, and other basic elements of function and process. Moreover, additional case studies are needed to determine whether the same is true for compounds including potent endogenous hormones. However, available data on BPA indicate that complete removal from wastewater effluent constitutes an inefficient method of exposure prevention when compared to reducing other sources of exposure.

This is especially true in light of the high expense required to treat wastewater to advanced levels. Israel is already investing in treating its wastewater to high levels, but the PA has limited resources that must allocated to high priority needs at the expense of lower priorities. Thus, it is particularly important for the PA to understand the cost of tertiary wastewater treatment relative to the gains, and in consideration of other alternative uses for these funds. This is discussed in the next section below.

Cost-Effectiveness Analysis

The second analysis conducted based the chemical results was a cost-effectiveness analysis (CEA) for the West Bank. This CEA was intended to determine whether it would be better to maintain primary and secondary wastewater treatment facilities in the West Bank which remove most of the EDCs from wastewater, or to pay for tertiary treatment levels which remove slightly more EDCs from wastewater. A CEA compares mutually exclusive scenarios to assess their effectiveness per dollar spent. The ratio of a scenario's total cost to its quantified effectiveness is compared; the cost is divided by effectiveness to procure a cost-per-unit effectiveness measure for each option. The smallest ratio indicates the greatest effectiveness at the lowest cost [103]. A CEA cannot dictate if an investment is or is not worthwhile because it does not consider any non-market or environmental variables. However a CEA can compare between the effectiveness of investment choices based on a comparison of cost and incremental gain [104].

For this West Bank CEA, one primary municipal WWTP (Tul Karem) and two secondary (Al Bireh and West Nablus) were used as a case study. As no tertiary level WWTPs exist in the West Bank, two from Israel (Hod Hasharon and Raanana) were used as a surrogate since most of the relevant characteristics (local materials, electricity and wages) are comparable across the border [103]. For the CEA effectiveness measure, estrone and estriol removal rates were calculated by comparing concentrations in samples of wastewater influent and effluent at the WWTPs of different treatment levels (see Figure 2 below). These two compounds were chosen for the case study as they contribute the most to the water's estrogenicity, and because there was adequate data available for both. Approximately a ten-fold increase in removal rates was observed between primary and secondary treatment levels, but only a slight increase of 2% between secondary and tertiary treatment levels. Yet attaining such a modest incremental improvement between secondary and tertiary treatment doubled the total cost of WWTP building and operations.

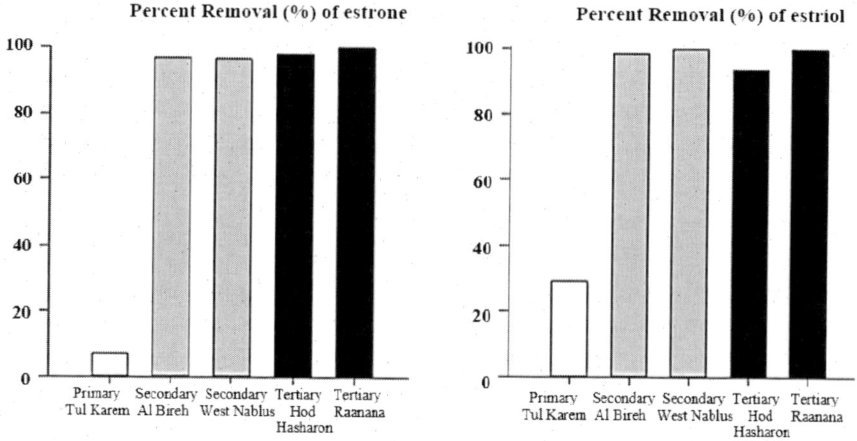

Adapted from Gordon-Kirsch [103].

Figure 2. Removal Rates at Each WWTP.

To calculate the cost values used in this CEA, the start-up costs of building and of operating a plant over time were used (See Table 5 below). When a large secondary WWTP in the West Bank (West Nablus) is compared to a somewhat smaller tertiary WWTP in Israel (Hod-Hasharon), associated costs for the secondary WWTP is lower across the board: 25% less for building costs, 26% less for yearly operations, and 30% less for the combined total (See Table 6 below).

It is important to consider size and capacity of WWTPs prior to drawing conclusions from cost comparisons of wastewater treatment levels. When cost was calculated per cubic meter of treated water and amortized (where initial costs are spread out over the life of the WWTP), secondary WWTPs are still less expensive at 2.3 NIS per cubic meter of wastewater (amortized) than tertiary WWTPs at 2.43 NIS per cubic meter of wastewater (amortized) when they are performing at full capacity, albeit by a smaller margin of 5% [103]. Using these effectiveness and cost measures, the Cost-Effectiveness (CE) ratio was calculated for secondary and tertiary WWTPs with costs including start-up and operation. Secondary treatment had a significantly lower CE ratio (See Table 7 below) confirming that this constitutes the optimal wastewater management strategy in the West Bank with regards to estriol and estrone removal [103].

Table 5. CEA Case Study WWTP Inflows, Startup Costs, and Operational Costs (Adapted from Gordon-Kirsch [103])

WWTP	Treatment	Daily average inflow (m^3/d)	Building Costs (NIS)	Yearly Operations + Maintenance (NIS/yr)
Tul Karem	Primary	~5,000 (cap 13,000)	2.302 million	120,000
Al Bireh	Secondary	~6,000 (cap 5,000)	34.72 million	1.88 million
West Nablus	Secondary	~9,000 (cap 15,000)	104.64 million	2.424 million (below inflow cap)
Hod Hasharon-Kfar Saba	Tertiary	25,895	138 million	14 million
Raanana	Tertiary	12,301	ND*	ND*

* ND = no data. Raanana WWTP provided no operational costs, so the individual cost of treatment per unit was used.

Table 6. Comparing Building, Operating, and Total Costs of CEA Secondary and Tertiary WWTPs in New Israeli Shekels (NIS) and US dollars (USD)

	Unit	Building Cost	Yearly Operational Cost	Total Cost
Secondary WWTP	Million NIS	104.64	2.42	107.06
	Million USD	27.35	.63	27.98
Tertiary WWTP	Million NIS	138.00	14.00	152.00
	Million USD	36.09	3.66	39.75
Savings of Secondary vs. Tertiary WWTP		25%	26%	30%

Table 7. Cost Effectiveness Ratio Values for Estrone and Estriol (Adapted from Gordon-Kirsch [103])

	Estrone	Estriol
Al Bireh Secondary Treatment	1.93	1.90
West Nablus Secondary Treatment	2.03	1.96
Hod Hasharon Tertiary Treatment	2.46	2.58
Raanana Tertiary Treatment	ND*	ND*

*ND = No data.

The CEA presented here indicates that investing in upgrading WWTPs to tertiary levels in the West Bank would not be an optimal strategy in terms of cost per unit removal of estriol and estrone. The vast majority of chemicals and any associated human health risks that they might pose by disrupting the endocrine system are eliminated when secondary level WWTPs are running smoothly, producing water safe enough to reuse for producing many agricultural products that are not eaten raw (e.g., cotton and fodder) [105, 106]. In fact, some farmers in Israel argue that wastewater treated to secondary levels is better than water treated to tertiary levels as it contains more nutrients and saves money on fertilizer [107, 108]. In light of the Palestinian government's limited resources in the face of many competing priorities, money saved by building WWTPs to secondary instead of more expensive tertiary levels could be utilized to expand water and sanitation services and sewage infrastructure in general producing far greater benefits to human and environmental well-being.

In cases where international donors are willing to fund the building costs of water infrastructure projects in the PA [109], there are several reasons why it would be wise for the Palestinian government to advocate for secondary WWTPs in lieu of tertiary level technologies at this time. First, the former cost less to build which will allow donors to stretch their budget to additional sanitation initiatives. For instance, surplus funding could go towards reducing the vast number of Palestinians who do not receive basic sewage collection and treatment services at all. Supplemental irrigation projects could also be funded with this money to help make use of treated wastewater (which is currently an under-tapped resource in the Palestinian territories), or to increasing local human capacity through training of engineers and other water management professionals.

Second, if outside funding covered the up-front capital costs of building WWTPs, the Palestinian government would be able to operate double the number of secondary WWTPs for the same money as they would spend to operate one tertiary WWTP, or it could operate one secondary WWTP for twice as long. This could double the number of people that would receive treatment of their sewage for the same cost. Building a secondary WWTP would help the PA make the most use of potential water resources in the face of growing scarcity due to climate change and a growing population [57]. Such a strategy would help decrease present Palestinian dependence on purchases from external sources such as *Mekorot*, the National Israeli water utility, or from private delivery trucks [57].

There is a real life example underway as plans to build a secondary WWTP in the Hebron Governance of the West Bank move forward. This WWTP will serve an estimated 104,000 people [103], and will treat sewage coming from 80% of the municipality. The plan involves four phases:

1) Building the WWTP to serve the immediate needs of treating 80% of the existing sewage;
2) constructing an agricultural irrigation system for reuse of the treated effluent;
3) building a network of sewers to capture the remaining 20% of the municipality which is unsewered; and
4) capacity building among communities in the area [57].

Financing the first phase of building the WWTP with a capacity to treat 15,000 CM/day of wastewater would cost an estimated 61.65 million USD, and phases two through four will cost a total of approximately 40 million USD. If the finding of the CEA is correct—building a tertiary WWTP can cost double the price of building a secondary WWTP— the anticipated savings from building a secondary treatment level WWTP in Hebron could be used to implement phases two through four. Alternatively, the extra money could be used to build another secondary WWTP that treats an additional 15,000 CM/day, easily covering the wastewater treatment needs of a medium sized city, or five smaller towns [110].

Operating twice the number of secondary WWTPs could also double the environmental gains by reducing the untreated wastewater that is presently released into the environment carrying pathogens and contaminants including EDCs. After treatment, this water could instead be used to grow crops and ameliorate the pervasive water scarcity in the West Bank. Already, a lack of wastewater treatment has contributed to extensive contamination of aquifers shared by Israel and the PA [111–114]. Studies have demonstrated that continuous contamination of the Eastern Mountain aquifer providing water to the West Bank would seriously undermine its future ability to provide any safe water [57].

In Israel, the considerations driving wastewater treatment decision were fundamentally different from this CEA conducted for the West Bank for two reasons. First, Israel has already overcome most of the basic water quality and water access challenge that are still the top problems in the PA. Second, even though irrigation with recycled water has contributed emerging contaminants in Israeli soil and groundwater, other well-known contaminants were the main concern. While secondary level WWTPs remove organic load as well as pathogens, they do not address boron, brines, or heavy metals which can contaminate soil [108]. These are released in effluent used for irrigation and can quickly build up to levels that threaten groundwater, soil, and crops. Consequently, the upgrade to tertiary treatment was first pursued in Israel to reduce levels of nutrients (nitrogen and phosphorous) in wastewater effluent; EDC removal was an unintended benefit. Wastewater treated to this level can expand the safe uses available for treated wastewater while avoiding nutrient buildup.

To finance such an investment in Israel, local authorities are usually responsible for operations and maintenance costs of WWTPs, while the initial costs of building or upgrading a WWTP are generally paid for by the Israeli national government. Israel's Water Authority has a "Master Plan" in place that estimates a cost of 150 million USD for building and upgrading WWTPs in Israel between 2015 and 2019 to meet national goals, and a cost of 673 million USD for treatment of wastewater [34]. Since then, an independent analysis estimated that the cost of upgrading Israeli treatment facilities to meet new standards would actually be 228 million USD with an annual cost of 33.23 million USD for operations and maintenance (132.92 million USD over 4 years) [115], which totals 360.92 million USD.

RESULTS

Israel: Advanced Wastewater Treatment for EDC Removal vs. Other Water Recycling Risks

In Israel, the combined estrogenicity of the five EDCs measured in treated wastewater was found to be approximately .617 ng/L EEQs, but it is important to put this number into context. Comparing daily EEQ doses from recycled wastewater calculated in this study to oral exposure from other sources, two orders of magnitude more estrogen was consumed via food and pharmaceuticals than through recycled wastewater [116], and four orders of magnitude more is consumed taking oral contraceptives [117]. EDC levels in the region's treated wastewater as a whole were found to be within global averages, and concentrations in Israel's treated wastewater were found to be below global averages [83]. See Figure 3 which compares concentrations in Israel to those from a combination of global studies [90, 97, 118–144].

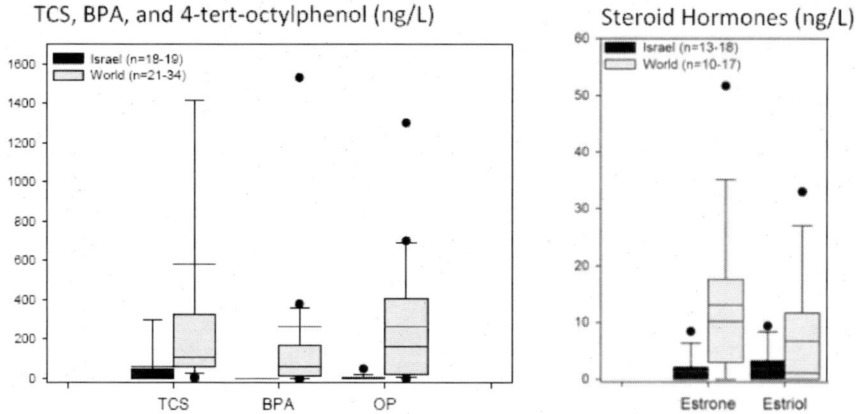

Box edges at 25th and 75th percentiles, error bars at 10th and 90th percentiles, red lines at averages, black lines at medians, and (•) at extreme values.

Adapted from Godinger [83].

Figure 3. EDCs Concentrations in Israel and Globally.

Until recently, only concentrations greater than 1 ng/L EEQ from in vitro assays were thought to be associated with adverse effects on individuals in vivo. Today, safe EEQs are believed to be lower: between 0.1 to 0.4 ng/L for wildlife [145], suggesting that levels of estrogenicity in Israel's treated wastewater effluent may threaten ecological health. Health effects of BPA exposure via recycled wastewater were only noteworthy where there was usually extreme exposure and/or preexisting semen conditions. This merits further study, both to confirm these findings and expand them through case studies on other EDC's and other health endpoints. However, the case study of human health risk from EDCs based on a BPA and fertility data indicates that risk to the general population from EDC exposure via recycled wastewater is relatively minimal.

PA: Benefits of Tertiary WWTP vs. Alternate Spending Plans

In treated Palestinian wastewater, the total estrogenicity based on the average concentrations of five EDCs was found to be 0.558 ng/L EEQs which is below levels in Israel, and very close to meeting the most stringent safety threshold estimates. Endogenous hormone levels in the West Bank's treated effluent are within the of distribution of global averages (see Figure 4 below) [84]. This should be confirmed by continued sampling in the PA locations included in this study to determine whether this situation is changing, and to reflect changes in water resource management and reuse. For now, this serves as the best data-driven indicator available that secondary treatment is sufficient to protect human health, even assuming expanded recycling of effluents there in the future.

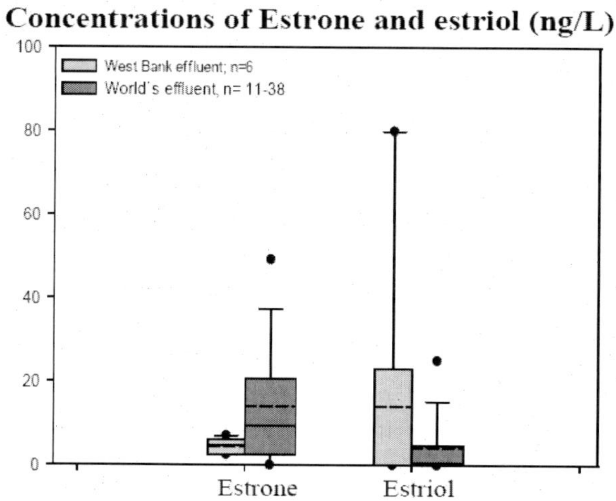

Box edges at 25th and 75th percentiles, error bars at 5th and 95th percentiles, dashed lines at averages, black lines at medians, and (•) at outliers.
Adapted from Odeh [84].

Figure 4. Concentrations of Estrone and Estriol in West Bank's Treated Wastewaters vs. World's.

DISCUSSION

Recommendations for Israel

The chemical, exposure risk, and cost analyses have helped clarify the best direction for a wastewater treatment strategy in Israel and similar developed nations. None of the challenges associated with treated wastewater necessitate abandoning the practice of sewage recycling in Israel. In fact, the country should complete the upgrading of its existing network of WWTPs to tertiary wastewater treatment in order to protect the soil and water from non-EDC threats (e.g., salinization, excess nutrients and other organic micro-pollutants). Tertiary and advanced treatment should continue to be an important priority and critical part of a strategy to open up the possibility of unlimited agricultural uses or even direct ingestion of treated wastewater without increasing risk to human health, but is not essential in Israel from the perspective of addressing human health risks from EDC exposure at this time. Advancing beyond tertiary levels to additional expensive advanced treatments to address EDC exposure at this stage would be premature, however EDC exposure can be addressed through other more targeted and comprehensive mitigation options.

Management of EDC exposure goes well beyond wastewater treatment to sources such as food, personal care products, and pharmaceuticals that should be addressed as a package. The Israeli government should begin to confront overall EDC exposure, but wastewater reuse should not be the central focus of new regulation. Advanced nations like Israel that have already met citizens' basic needs and wish to pursue EDC exposure reduction are advised to focus on other more significant and straightforward exposure sources such as foods containing soy, food containers such as cans and plastic packages, and polycarbonate cooking tools. Source reduction efforts can include take-back programs for unused pharmaceuticals, fines for manufacturing domestic personal care products and foods containing prominent estrogenic EDCs, and/or subsidies for development of safer replacement compounds and products [146–148].

To ensure that EDC concentrations in recycled wastewater do not contribute more to human exposure in the future, water should be checked for concentration levels of common EDCs, and for estrogenicity overall Water for drinking and bathing should be monitored for EDCs, as well as applications such as effluent irrigation for water-fed produce, and groundwater which may be vulnerable [149]. The Israeli government would do well to ensure its EDC screening and testing procedures remain in line with global best practices over time. To reach its maximum potential, Israel's should establish a comprehensive basin-wide EDC monitoring policy with Palestinian input [150, 151]. The importance of such cooperation will be discussed further in the Conclusions section below.

Recommendations for PA

The analyses conducted in this study also clarify several realities that should help inform the PA operational decisions regarding the importance of EDCs as a water policy driver in developing countries. When the marginal improvement from tertiary wastewater treatment is considered in light of the relatively modest budgets available for infrastructure, upgrading to tertiary levels would not return an improvement in EDC reduction proportionate to the expense. The cost savings from foregoing tertiary treatment when building new WWTPs across Palestine could be utilized for other basic infrastructure and sanitation service needs that are still lacking including pipes for wastewater removal and fresh water distribution, schemes for irrigation with treated wastewater that adhere to international standards [152], and capacity building projects that teach locals to operate wastewater systems and utilize the products efficiently and sustainably.

This money could also be put towards addressing myriad non-EDC insults associated with wastewater reuse that pose the highest risk to human and environmental health: release of pathogens, and soil or plant accumulation of nutrients and metals. Today 20-25% of Palestinian households still lack reasonable wastewater collection and sewage treatment [6, 56]. The result of such low wastewater treatment capacity results in the situation currently found in the Palestinian territories; for lack of a better option, large portions of the sewage produced by the population is released untreated into the environment. Historically, the Palestinian people have borne the burden of this deficiency in their shared region, including periodic outbreaks of hepatitis, typhoid, and dysentery [153]. Even in recent decades, infants still die arguably preventative deaths from infectious disease associated with water and hygiene. Hospitalizations and unreported illnesses are also very common although less difficult to document [107]. Treating all sewage to secondary levels would provide treated water of reasonable quality to more people, and would benefit the natural environment on which they rely.

If the public policy objective were purely based on maximal reduction of all risks of endocrine disruption, then contaminant removal via tertiary treatment may make the most sense [103]. However considering Palestinian economic challenges, the limited wastewater management capacity and infrastructure of the PA at present, and the health risks posed by conventional contaminants, resources spent on this incremental improvement in the quality of treated water would be better utilized by the PA to address more pressing needs. Secondary treatment is thus a preferable choice, especially if accompanied by guidelines on safe use of such effluent for reuse. Over time secondary plants can easily be upgraded to tertiary levels once investments in water infrastructure and capacity have met basic needs. However, until the vast majority of sewage is collected and treated, Palestinian decision makers would be wise to invest in secondary WWTPs and other basic infrastructure and capacity building projects to protect public health and best serve the wellbeing of their citizens.

To this end, future Palestinian policy should proceed according to two iterative stages: expeditiously building WWTPs that contain primary and secondary treatment facilities to address the most urgent environmental and human health threats from traditional contaminants in untreated sewage; and eventually upgrading to tertiary treatment levels, but only after these basic needs are addressed. At the same time, policy and guidance encouraging safe and efficient effluent use must be provided to allow Palestinians to make the best use of the water they already have. As of 2011, more than 40% of the water supply was allocated for agricultural use (as opposed to domestic use) in the Palestinian territories [58]. Water managers should seek to lower current Palestinian limits on concentrations of contaminants to allow reuse of effluent in agriculture or aquifers [154]. Such effluent used safely would free up fresh water for drinking and other forms of direct consumption. Effluents must play a key role in sustainable water management and development if the Gaza Strip and West Bank are to provide citizens with enough clean water [155–157].

A timeline should be laid out for this development. Inter-Palestinian cooperation between its Ministry of Health, Ministry of Planning and International Cooperation, Environmental Quality Authority, and Legislative Council will be indispensable in coordinating enforcement of the laws and regulations relevant to water. These efforts should be supported by Israel through knowledge sharing and financial support as the benefits of this investment will be shared across the border.

Regional Cooperation

Israeli cooperation with the PA will be necessary to ensure that EDC environmental and health protection measures are effective. Under a "temporary" agreement that is almost two decades old, Israel and Palestine have had only modest success at joint management of their water, but surely not enough and not always in official state capacities [158]. In the past, animosities between the two governments have prevented the establishment and cooperative enforcement of basin-wide standards [150, 151]. Palestinians have been hesitant to sign a request for assistance for fear of weakening their negotiation position, and joint projects were blocked by similar fears regarding power positions on both sides [159]. In general, Israel has chosen to make unilateral water management decisions as the benefits have been high, including water independence and decreased vulnerability [160].

For a sustainable regional water policy to emerge, Israel and the PA should negotiate a way to "share the costs of transboundary wastewater treatment," and avoid unfair distribution of negative transboundary externalities with regards to water use [160, 161]. Positive steps in this direction could include Palestinians receiving the benefits of information sharing, capacity building, new wastewater treatment plants in Palestinian territories [162], and transfers of desalinated water as an emergency measure in the meantime [149, 163]. This cooperation could lead to the necessary basin-wide initiatives [162] and basin-wide approach to management [164]. Future initiatives must include curbing the amount of untreated sewage released into the environment. This will enable the region to both eliminate contamination from wastewater discharges, and expand its freshwater supply to achieve water security.

Conclusion

While Israel has largely overcome traditional hygiene and sanitation obstacles and can focus on emerging water quality challenges, the West Bank potentially faces both. The question answered by this study is whether EDCs should drive water policy in Israel and in the PA—specifically regarding the decision of whether to further upgrade wastewater treatment levels—based on the current state of knowledge. At this point in time, the answer is that upgrading wastewater treatment is *not* the best way to address the most urgent human health risks from raw wastewater in the Palestinian territories, nor from EDCs in Israel. Findings indicate that risk from traditional contaminants, not EDCs, should still be the driver for water policy and wastewater management in both countries, albeit for different reasons.

This study on EDCs in the region was intended not only to inform the water policies of developed and developing nations in general, and provides essential lessons regarding the fate of environments shared by groups with different economic capacities. Neighboring countries are likely to feel the effects of the water management strategies being implemented by their neighbors, regardless of their own infrastructure and services. They will share the consequences of damage to the region's natural environment, regardless of a political and socio-economic separation. Thus, when an area of economic disparity faces environmental contamination, the lesson learned from this case study is to strive for cooperation.

Such cross border environmental and public health agreements have been successful in other areas of economic disparity including the US and Mexico [165], between eastern and western Europe [166], in a few cases in Haiti and the Dominican Republic [167], and elsewhere [168]. In Israel and the West Bank, the ideal solution would be a unified basin-wide policy with agreements to jointly managing wastewater so that populations on both sides are protected. Considering the vastly different political and economic reality faced by Israel and the West Bank, such an initiative would be greatly facilitated by knowledge transfers and training from Israel to the PA. It would also help collaboration if join projects were funded proportionately by each government based on financial capacity. This will be especially challenging to achieve at a national level with governments in political conflict zones, but in cases where government are unwilling to negotiate, non-governmental organizations and academic institutions can step in. Research projects such as this are a foundational first step.

Improving wastewater treatment should be a shared Israeli and Palestinian goal in order reduce all types of contamination in their shared water basin, and to protect their common environment. It is clear from the case study in Israel and the PA that EDCs can be found in the aquatic environment of the entire region, so monitoring EDC levels should become part of both governments' regular water management protocol in case the situation should change in the future. Without significant efforts on both sides of the border, Israelis and Palestinians will all continue to face both traditional and emerging contaminants in wastewater released to the environment. The land and water resources of the region until now have provided food and water to sustain both peoples, but may be unable to do so in the future without interventions. Developing and developed nations can both benefit from this lesson, especially in areas where both economic realities inhabit a shared environment.

References

[1] Israel Water Authority, Water Sector in Israel: *IWRM Model,* (2015).

[2] A. Tal, Enduring Technological Optimism: Zionism's Environmental Ethic and its Influence on Israel's Environmental History, *Environ. Hist. Durh. N.* C. 13 (2008) 275–305.

[23] A. A. Kortenkamp, O. Martin, M. Faust, R. Evans, R. Mckinlay, F. Orton, et al., State of the art assessment of endocrine disrupters, 2011. http://scholar.google.com/scholar?hl=en&btnG=Search&q=intitle:STATE+OF+THE+ART+ASSESSMENT+OF+ENDOCRINE+DISRUPTERS+Final+Report#5.

[24] R. T. Zoeller, T. R. Brown, L. L. Doan, A. C. Gore, N. E. Skakkebaek, A. M. Soto, et al., Endocrine-Disrupting Chemicals and Public Health Protection: A Statement of Principles from The Endocrine Society, *Endocrinology.* 153 (2012) 4097–4110. doi:10.1210/en.2012-1422.

[25] E. Diamanti-Kandarakis, J.-P. Bourguignon, L. C. Giudice, R. Hauser, G. S. Prins, A. M. Soto, et al., Endocrine-Disrupting Chemicals: An Endocrine Society Scientific Statement, *Endocr. Rev.* 30 (2009) 293–342. doi:10.1210/er.2009-0002.

[26] C. M. Markey, M. a. Coombs, C. Sonnenschein, A. M. Soto, Mammalian development in a changing environment: exposure to endocrine disruptors reveals the developmental plasticity of steroid-hormone target organs, *Evol. Dev.* 5 (2003) 67–75. doi:10.1046/j.1525-142X.2003. 03011.x.

[27] C. M. Markey, P. R. Wadia, B. S. Rubin, C. Sonnenschein, A. M. Soto, Long-term effects of fetal exposure to low doses of the xenoestrogen bisphenol-A in the female mouse genital tract, *Biol. Reprod.* 72 (2005) 1344–1351. doi:10.1095/biolreprod.104.036301.

[28] G. Schönfelder, K. Friedrich, M. Paul, I. Chahoud, Developmental Effects of Prenatal Exposure to Bisphenol A on the Uterus of Rat Offspring, *Neoplasia.* 6 (2004) 584–594. doi:10.1593/neo.04217.

[29] L. N. Vandenberg, T. Colborn, T. B. Hayes, J. J. Heindel, D. R. Jacobs, D.-H. Lee, et al., Regulatory decisions on endocrine disrupting chemicals should be based on the principles of endocrinology, *Reprod. Toxicol.* 38 (2013) 1–15. doi:10.1016/j.reprotox.2013.02.002.

[30] O. V Martin, J.N. Lester, N. Voulvoulis, A.R. Boobis, Human health and endocrine disruption: A simple multicriteria framework for the qualitative assessment of end point-specific risks in a context of scientific uncertainty, *Toxicol. Sci.* 98 (2007) 332–347. doi:10.1093/toxsci/kfm008.

[31] D. Gee, Establishing evidence for early action: The prevention of reproductive and developmental harm, *Basic Clin. Pharmacol. Toxicol.* 102 (2008) 257–266. doi:10.1111/j.1742-7843.2008.00207.x.

[32] J. M. Muller, Minutes of the expert meeting on endocrine disruptors, (2013) 1–5. ec.europa.eu/commission_2010-2014/president/chief-

scientific-adviser/documents/minutes_endocrine_disruptors_meeting_241013_final.pdf.

[33] Z. Rinat, Israeli Wastewater Policy Continues to Pay Off, Ha'aretz. (2015). http://www.haaretz.com/life/nature-environment/.premium-1.648332.

[34] Israel Water Authority, Master Plan for the National Water Sector: Main Points of the Policy Paper, (2012). http://www.water.gov.il/Hebrew/ProfessionalInfoAndData/2012/09-Israel-Water-Sector-Master-Plan-2050.pdf.

[35] Y. Inbar, New Standards for Treated Wastewater Reuse in Israel, in: M. Zaidi (Ed.), Wastewater Reuse–Risk Assessment, *Decis. Environ. Secur., Springer Netherlands, Dordrecht,* 2007: pp. 291–296. doi:10.1007/978-1-4020-6027-4_28.

[36] S. Weber, S. Khan, J. Hollender, Human risk assessment of organic contaminants in reclaimed wastewater used for irrigation, *Desalination.* 187 (2006) 53–64. doi:10.1016/j.desal.2005.04.067.

[37] P. Anderson, N. Denslow, J. E. Drewes, A. Olivieri, D. Schlenk, S. Snyder, Final Report, Monitoring Strategies for Chemicals of Emerging Concern (CECs) in Recycled Water–Recommendations of a Science Advisory Panel Panel Members, Sacramento, CA, 2010. http://www.swrcb.ca.gov/water_issues/programs/water_recycling_policy/docs/cec_monitoring_rpt.pdf.

[38] S. A. Snyder, D. L. Villeneuve, E. M. Snyder, J. P. Giesy, Identification and quantification of estrogen receptor agonists in wastewater effluents, *Environ. Sci. Technol.* 35 (2001) 3620–3625. doi:10.1021/es001254n.

[39] C. H. Huang, D. L. Sedlak, Analysis of estrogenic hormones in municipal wastewater effluent and surface water using enzyme-linked immunosorbent assay and gas chromatography/tandem mass spectrometry., *Environ. Toxicol. Chem.* 20 (2001) 133–139. doi:10.1897/1551-5028(2001)020<0133:AOEHIM>2.0.CO;2.

[40] A. Joss, H. Andersen, T. Ternes, P. R. Richle, H. Siegrist, Removal of estrogens in municipal wastewater treatment under aerobic and anaerobic conditions: Consequences for plant optimization, *Environ. Sci. Technol.* 38 (2004) 3047–3055. doi:10.1021/es0351488.

[41] B. L. L. Tan, D. W. Hawker, J. F. Müller, F. D. L. Leusch, L. A. Tremblay, H. F. Chapman, Comprehensive study of endocrine disrupting compounds using grab and passive sampling at selected wastewater treatment plants in South East Queensland, Australia, *Environ. Int.* 33

(2007) 654–669. http://www.sciencedirect.com/science/article/B6V7X-4N55THX-1/2/e6333caee3c630f858329bc36cea3e19.

[42] G. G. Ying, R. S. Kookana, A. Kumar, M. Mortimer, Occurrence and implications of estrogens and xenoestrogens in sewage effluents and receiving waters from South East Queensland, *Sci. Total Environ.* 407 (2009) 5147–5155. doi:10.1016/j.scitotenv.2009.06.002.

[43] N. Ratola, A. Cincinelli, A. Alves, A. Katsoyiannis, Occurrence of organic microcontaminants in the wastewater treatment process. A mini review, *J. Hazard. Mater.* 239-240 (2012) 1–18. doi:10.1016/j.jhazmat.2012.05.040.

[44] Y. Lee, B.I. Escher, U. von Gunten, Efficient Removal of Estrogenic Activity during Oxidative Treatment of Waters Containing Steroid Estrogens, *Environ. Sci. Technol.* 42 (2008) 6333–6339. doi:10.1021/es7023302.

[45] D. Katz, A. Tal, Rehabilitating Israel's Streams and Rivers, in: *Water Policy Isr.*, 2013: pp. 65–81. doi:10.1007/978-94-007-5911-4_5.

[46] A. C. Johnson, J. P. Sumpter, Removal of Endocrine-Disrupting Chemicals in Activated Sludge Treatment Works, *Environ. Sci. Technol.* 35 (2001) 4697–4703. doi:10.1021/es010171j.

[47] M. L. Janex-Habibi, A. Huyard, M. Esperanza, A. Bruchet, Reduction of endocrine disruptor emissions in the environment: The benefit of wastewater treatment, *Water Res.* 43 (2009) 1565–1576. doi:10.1016/j.watres.2008.12.051.

[48] J. H. Lee, J. L. Zhou, S. D. Kim, Effects of biodegradation and sorption by humic acid on the estrogenicity of 17??-estradiol, *Chemosphere.* 85 (2011) 1383–1389. doi:10.1016/j.chemosphere.2011.08.003.

[49] Z. H. Liu, J. A. Ogejo, A. Pruden, K. F. Knowlton, Occurrence, fate and removal of synthetic oral contraceptives (SOCs) in the natural environment: A review, *Sci. Total Environ.* 409 (2011) 5149–5161. doi:10.1016/j.scitotenv.2011.08.047.

[50] E. R. V Dickenson, S. a. Snyder, D. L. Sedlak, J. E. Drewes, Indicator compounds for assessment of wastewater effluent contributions to flow and water quality, *Water Res.* 45 (2011) 1199–1212. doi:10.1016/j.watres.2010.11.012.

[51] B. Chefetz, T. Mualem, J. Ben-Ari, Sorption and mobility of pharmaceutical compounds in soil irrigated with reclaimed wastewater, *Chemosphere.* 73 (2008) 1335–1343. doi:10.1016/j.chemosphere.2008.06.070.

[52] T. Malchi, Y. Maor, G. Tadmor, M. Shenker, B. Chefetz, Irrigation of Root Vegetables with Treated Wastewater: Evaluating Uptake of Pharmaceuticals and the Associated Human Health Risks, *Environ. Sci. Technol.* 48 (2014) 9325–9333. doi:10.1021/es5017894.

[53] M. Shenker, D. Harush, J. Ben-Ari, B. Chefetz, Uptake of carbamazepine by cucumber plants - A case study related to irrigation with reclaimed wastewater, *Chemosphere.* 82 (2011) 905–910. doi:10.1016/j.chemosphere.2010.10.052.

[54] OECD-- Organisation for Economic Co-operation and Development, Water Resources Allocation: Sharing Risks and Opportunities, 2015. doi:10.1787/9789264229631.

[55] R. Al-Sa'ed, N. Tomaleh, Performance Evaluation of a Full-Scale Extended Aeration System in Al-Bireh City, Palestine, CLEAN - Soil, Air, Water. 40 (2012) 1250–1256. doi:10.1002/clen.201000095.

[56] PWA-- Palestinian Water Authority, Annual water status report 2011, 2012.

[57] World Bank, Hebron Regional Wastewater Management Project - Phase 1 Project Information Document (PID), 2014. http://www-wds.worldbank.org/external/default/WDSContentServer/WDSP/MNA/2014/10/01/090224b08279b472/1_0/Rendered/PDF/Project0Inform0000Phase01000P117449.pdf.

[58] PWA-- Palestinian Water Authority, Water Supply Report 2010, Al-Baloua, West Bank; Al-Rimal, Gaza, 2012. http://www.pwa.ps/Portals/_PWA/Supply water report to print%282%29.pdf (05/06/13).

[59] A. Yasin, Personal interview conducted by Gordon-Kirsch for 2015 masters thesis, (2014).

[60] Z. Tagar, V. Qumsieh, A Seeping Time Bomb: Pollution of the Mountain Aquifer by Solid Waste, Tel Aviv and Bethlehem, 2006. http://foeme.org/uploads/publications_publ59_1.pdf.

[61] S. B. Megdal, R. G. Varady, S. Eden, eds., Shared Borders, Shared Waters: Israeli-Palestinian and Colorado River Basin Water Challenges, CRC Press, London, England, 2013.

[62] Metcalf & Eddy Inc., Water Engineering: Treatment and Reuse, McGraw-Hill, Boston, MA, 2003.

[63] C. Chodick, M. Huerta, R. Balicer, N. Davidovitch, I. Grotto, Secular trends in age at menarche, smoking, and oral contraceptive use among Israeli girls, Prev Chronic Dis. 2 (2005) A12. http://www.cdc.gov/pcd/.

[64] S. Flashe-Luzzatti, C. Weil, V. Shalev, T. Oron, G. Chodick, Long-Term Secular Trends in the Age at Menarche in Israel: A Systematic Literature

Review and Pooled Analysis, *Horm. Res. Paediatr.* (2014). doi:10.1159/000357444.

[65] T. Sella, G. Chodick, E. Lunenfeld, V. Shalev, Further evidence on the high prevalence of male factor infertility diagnosis in Israel., *Isr. Med. Assoc. J.* 13 (2011) 386.

[66] J. Farhi, A. Ben-Haroush, Distribution of causes of infertility in patients attending primary fertility clinics in Israel, *Isr. Med. Assoc. J.* 13 (2011) 51–54.

[67] J. Dor, R. Homburg, E. Rabau, An evaluation of etiologic factors and therapy in 665 infertile couples., *Fertil. Steril.* 28 (1977) 718–722. doi:10.1097/00006254-197801000-00020.

[68] M. Bar-Chana, Cancer Incidence Tables, (2010). http://www.health.gov.il/download/sartan/trends/Testis.pdf.

[69] M. Almagor, I. Ivnitzki, H. Yaffe, M. Baras, Changes in Semen Quality in Jerusalem Between 1990 AND 2000: A Cross-Sectional and Longitudinal Study, *Arch. Androl.* 49 (2003) 139–144. doi:10.1080/713828098.

[70] A. Benshushan, O. Shoshani, O. Paltiel, J. G. Schenker, A. Lewin, Is there really a decrease in sperm parameters among healthy young men? A survey of sperm donations during 15 years, *J. Assist. Reprod. Genet.* 14 (1997) 347–353. doi:10.1007/BF02765840.

[71] R. Haimov-Kochman, R. Har-Nir, E. Ein-Mor, V. Ben-Shoshan, C. Greenfield, I. Eldar, et al., Is the quality of donated semen deteriorating? Findings from a 15 year longitudinal analysis of weekly sperm samples, *Isr. Med. Assoc. J.* 14 (2012) 372–377.

[72] A. Tal, Management of transboundary wastewater discharges Wastewater management in urban communities along the Israeli-Palestinian "border," in: S. B. Megdal, R. G. Varady, S. Eden (Eds.), Shar. Borders, Shar. Waters, *CRC Press, Boca Raton, FL*, 2012: pp. 221–232.

[73] T. Berman, H. Levine, R. Gamzu, I. Grotto, Trends in reproductive health in Israel: implications for environmental health policy., *Isr. J. Health Policy Res.* 1 (2012) 1–34. doi:10.1186/2045-4015-1-34.

[74] L. Muszkat, First Results of Research that Examined Ground Water Point to Pollution by Hazardous Organic Materials, Biosphere. 7 (1988) 15.

[75] L. Muszkat, Large Scale Contamination of Deep Groundwaters by Organic Pollutants, *Adv. Mass Spectrom.* 11B (1990) 1628.

[76] L. Muszkat, Penetration of pesticides and industrial organics into the depth of soil and groundwater, *Biochem Physiol.* 22 (1993) 487–99.

[77] D. Avisar, Y. Lester, D. Ronen, Sulfamethoxazole Contamination of a Deep Phreatic Aquifer, *Sci. Total Environ.* 407 (2009) 4278–4282.

[78] A. Lamm, A. Mayer, I. Gozlan, D. Avisar, Detection of Amoxicillin-Diketopiperazine- 2″, 5″ in Wastewater Samples, *J. Of. Environ. Sci. And. Heal. Part A*, 44 (2009) 1512–1514.

[79] G. Gasser, M. Rona, A. Voloshenko, R. Shelkov, N. Tal, I. Pankratov, et al., Quantitative Evaluation of Tracers for Quantification of Wastewater Contamination of Potable Water Sources, *Environ. Sci. Technol.* 44 (2010) 3919–3925. doi:10.1021/es100604c.

[80] K. Barel-Cohen, L. S. Shore, M. Shemesh, A. Wenzel, J. Mueller, N. Kronfeld-Schor, Monitoring of natural and synthetic hormones in a polluted river., *J. Environ. Manage.* 78 (2006) 16–23. doi:10.1016/j.jenvman.2005.04.006.

[81] O. Lev, Micro pollutants in effluents receiving various levels of treatment -- .2102 ,melasureJ ,מיקרומזהמים בקולחים בדרגות טיהור שונות

[82] D. Shargil, Z. Gerstl, P. Fine, I. Nitsan, D. Kurtzman, Impact of biosolids and wastewater effluent application to agricultural land on steroidal hormone content in lettuce plants, *Sci. Total Environ.* 505 (2015) 357–366. doi:10.1016/j.scitotenv.2014.09.100.

[83] T. Godinger, The Occurrence and Fate of Endocrine Disrupting Compounds in Wastewater Treatment Plants in Israel, Ben-Gurion University of the Negev, 2015.

[84] W. Odeh, Occurrence and Fate of Endocrine Disruptors in Wastewater and the Receiving Aquatic Environments in the West Bank of Palestine, Ben-Gurion University of the Negev, n.d.

[85] T. Manickum, W. John, Occurrence, fate and environmental risk assessment of endocrine disrupting compounds at the wastewater treatment works in Pietermaritzburg (South Africa), *Sci. Total Environ.* 468-469 (2014) 584–597. doi:10.1016/j.scitotenv.2013.08.041.

[86] T. Vega-Morales, Z. Sosa-Ferrera, J. J. Santana-Rodríguez, Evaluation of the Presence of Endocrine-Disrupting Compounds in Dissolved and Solid Wastewater Treatment Plant Samples of Gran Canaria Island (Spain), *Biomed Res. Int.* 2013 (2013) 1–15. doi:10.1155/2013/790570.

[87] B. Roig, W. Mnif, A. I. Hadj Hassine, I. Zidi, S. Bayle, A. Bartegi, et al., Endocrine Disrupting Chemicals and Human Health Risk Assessment: A Critical Review, *Crit. Rev. Environ. Sci. Technol.* 43 (2013) 2297–2351. doi:10.1080/10643389.2012.672076.

[88] R. Brix, C. Postigo, S. González, M. Villagrasa, A. Navarro, M. Kuster, et al., Analysis and occurrence of alkylphenolic compounds and estrogens in a European river basin and an evaluation of their importance as priority pollutants, *Anal. Bioanal. Chem.* 396 (2010) 1301–1309. doi:10.1007/s00216-009-3358-8.

[89] J. Blavier, G. Songulashvili, F. Debaste, Identification, test and comparison of analysis methods of urban wastewaters estrogenic activity, *Univ. Libr. Bruxelles.* (2013) 1–50.

[90] J. E. Drewes, J. Hemming, S.J. Ladenburger, J. Schauer, W. Sonzogni, An Assessment of Endocrine Disrupting Activity Changes during Wastewater Treatment through the Use of Bioassays and Chemical Measurements, *Water Environ. Res.* 77 (2005) 12–23. doi:10.2175/106143005X41573.

[91] S. Jobling, J. P. Sumpter, Detergent components in sewage effluent are weakly oestrogenic to fish: An in vitro study using rainbow trout (Oncorhynchus mykiss) hepatocytes, *Aquat. Toxicol.* 27 (1993) 361–372. doi:10.1016/0166-445X(93)90064-8.

[92] R. Zoeller, Å. Bergman, G. Becher, P. Bjerregaard, R. Bornman, I. Brandt, et al., A path forward in the debate over health impacts of endocrine disrupting chemicals, *Environ. Heal.* 13 (2014) 118. doi:10.1186/1476-069X-13-118.

[93] W. V. Welshons, K. a. Thayer, B. M. Judy, J. a. Taylor, E. M. Curran, F. S. vom Saal, Large Effects from Small Exposures. I. Mechanisms for Endocrine-Disrupting Chemicals with Estrogenic Activity, *Environ. Health Perspect.* 111 (2003) 994–1006. doi:10.1289/ehp.5494.

[94] C. Desbrow, E. J. Routledge, G. C. Brighty, J. P. Sumpter, M. Waldock, Identification of Estrogenic Chemicals in STW Effluent. 1. Chemical Fractionation and in Vitro Biological Screening, *Environ. Sci. Technol.* 32 (1998) 1549–1558. doi:10.1021/es9707973.

[95] T. P. Rodgers-Gray, S. Jobling, S. Morris, C. Kelly, S. Kirby, A. Janbakhsh, et al., Long-Term Temporal Changes in the Estrogenic Composition of Treated Sewage Effluent and Its Biological Effects on Fish, *Environ. Sci. Technol.* 34 (2000) 1521–1528. doi:10.1021/es991059c.

[96] H.-R. Aerni, B. Kobler, B. V. Rutishauser, F. E. Wettstein, R. Fischer, W. Giger, et al., Combined biological and chemical assessment of estrogenic activities in wastewater treatment plant effluents, *Anal. Bioanal. Chem.* 378 (2004) 688–696. doi:10.1007/s00216-003-2276-4.

[97] M. Auriol, Y. Filali-Meknassi, R. D. Tyagi, C. D. Adams, R. Y. Surampalli, Endocrine disrupting compounds removal from wastewater, a new challenge, *Process Biochem.* 41 (2006) 525–539. doi:10.1016/j.procbio.2005.09.017.

[98] V. Futran Fuhrman, A. Tal, S. Arnon, Why endocrine disrupting chemicals (EDCs) challenge traditional risk assessment and how to respond, *J. Hazard. Mater.* 286 (2015) 589–611. doi:10.1016/j.jhazmat.2014.12.012.

[99] EFSA Scientific Committee, Scientific Opinion on the hazard assessment of endocrine disruptors : Scientific criteria for identification of endocrine disruptors and appropriateness of existing test methods for assessing effects mediated by these substances on human health and the en, *EFSA J.* 11 (2013) 3132–3215. doi:10.2903/j.efsa.2013.3132.

[100] EFSA CEF Panel, Scientific Opinion on the risks to public health related to the presence of bisphenol A (BPA) in foodstuffs: Part I-- Exposure Assessment, *EFSA J.* 13 (2015) 1–397. doi:10.2903/j.efsa.2015.3978.

[101] D.-K. Li, Z. Zhou, M. Miao, Y. He, J. Wang, J. Ferber, et al., Urine bisphenol-A (BPA) level in relation to semen quality, *Fertil. Steril.* 95 (2011) 625–630.e4. doi:10.1016/j.fertnstert.2010.09.026.

[102] J.W. van der Steeg, P. Steures, M.J.C. Eijkemans, J.D. F. Habbema, P.G.A. Hompes, J.A.M. Kremer, et al., Role of semen analysis in subfertile couples, *Fertil. Steril.* 95 (2011) 1013–1019. doi:10.1016/j.fertnstert.2010.02.024.

[103] N. Gordon-Kirsch, Solutions for Palestinian Wastewater: A Cost Effectiveness Analysis of Wastewater Treatment Facilities in the Palestinian National Authority, Ben-Gurion University of the Negev, 2015.

[104] S. R. Cellini, J. E. Kee, Cost - Effectiveness and Cost - Benefit Analysis, in: Handb. *Pract. Progr. Eval.*, 2010: pp. 493–530.

[105] H. Shuval, International Seminar on Soil and Water Utilization, in: *Waste Water Util. Isr.*, Brookings, South Dakota, 1962: p. 40.

[106] D. Bixio, C. Thoeye, J. De Koning, D. Joksimovic, D. Savic, T. Wintgens, et al., Wastewater reuse in Europe, *Desalination.* 187 (2006) 89–101. doi:10.1016/j.desal.2005.04.070.

[107] A. Tal, From Development to Conservation: The Quantity and Quality of Israel's Water Resources, in: Pollut. a Promised L. *An Environ. Hist.* Isr., University of California Press, Berkley, CA, 2002.

[108] A. Tal, Rethinking the Sustainability of Israel's Irrigation Practices in the Drylands, in: Living-Land, Tudor Rose and United Nations

Convention to Combat Desertification, London, England, 2015. http://www.unccd.int/en/media-center/MediaNews/Pages/highlight detail.aspx?HighlightID=383.

[109] World Bank, Global Reach Map: West Bank and Gaza, (2015). http://maps.worldbank.org/p2e/mcmap/map.html?org=ibrd&level=count ry&code=GZ&title=West Bank and Gaza (accessed August 6, 2015).

[110] Palestinian Central Bureau of Statistics, Population, Housing and Establishment Census 2007: Census Final Results in The West Bank--Summary (Population and Housing), Ramallah, Palestine, 2008. http://www.pcbs.gov.ps/Portals/_PCBS/Downloads/book1487.pdf.

[111] EcoPeace Middle East, Protecting Groundwater, 2014. doi:10.1007/BF02929903.

[112] D. Avisar, J. Kronfeld, A. Siep Talma, Amelioration of groundwater nitrate contamination following installation of a central sewage system in two Israeli villages, *Environ. Geol.* 58 (2009) 515–520. doi:10.1007/s00254-008-1525-4.

[113] B. Borst, N. J. Mahmoud, N. P. van der Steen, P. N. L. Lens, A case study of urban water balancing in the partly sewered city of Nablus-East (Palestine) to study wastewater pollution loads and groundwater pollution, *Urban Water J.* 10 (2013) 434–446. doi:10.1080/1573062X. 2012.750373.

[114] S. Schmidt, T. Geyer, A. Marei, J. Guttman, M. Sauter, Quantification of long-term wastewater impacts on karst groundwater resources in a semi-arid environment by chloride mass balance methods, *J. Hydrol.* 502 (2013) 177–190. doi:10.1016/j.jhydrol.2013.08.009.

[115] D. Lavee, Is the upgrading of wastewater treatment facilities to meet more stringent standards economically justified: The case of Israel, *Water Resour.* 41 (2014) 564–573. doi:10.1134/S0097807814050042.

[116] I. R. Falconer, H. F. Chapman, M. R. Moore, G. Ranmuthugala, Endocrine-disrupting compounds: a review of their challenge to sustainable and safe water supply and water reuse., *Environ. Toxicol.* 21 (2006) 181–91. doi:10.1002/tox.20172.

[117] J. Pugh, M. Moore, Endocrine disruption. Australia's role in an international issue, Canberra, Australia, 1988. https://www.science.org.au/node/35560#.VdkVwU1RGP8.

[118] W. Körner, U. Bolz, W. Süßmuth, G. Hiller, W. Schuller, V. Hanf, et al., Input/output balance of estrogenic active compounds in a major municipal sewage plant in Germany, *Chemosphere.* 40 (2000) 1131–1142. doi:10.1016/S0045-6535(99)00362-8.

[119] T. Ternes, M. Stumpf, J. Mueller, K. Haberer, R.-D. Wilken, M. Servos, Behavior and occurrence of estrogens in municipal sewage treatment plants — I. Investigations in Germany, Canada and Brazil, *Sci. Total Environ.* 225 (1999) 81–90. doi:10.1016/S0048-9697(98)00334-9.

[120] M. R. Servos, D. T. Bennie, B. K. Burnison, A. Jurkovic, R. McInnis, T. Neheli, et al., Distribution of estrogens, 17β-estradiol and estrone, in Canadian municipal wastewater treatment plants, *Sci. Total Environ.* 336 (2005) 155–170. doi:10.1016/j.scitotenv.2004.05.025.

[121] D. E. Vidal-Dorsch, S. M. Bay, K. Maruya, S. a Snyder, R. a Trenholm, B. J. Vanderford, Contaminants of emerging concern in municipal wastewater effluents and marine receiving water., *Environ. Toxicol. Chem.* 31 (2012) 2674–82. doi:10.1002/etc.2004.

[122] H. Hamid, C. Eskicioglu, Fate of estrogenic hormones in wastewater and sludge treatment: A review of properties and analytical detection techniques in sludge matrix., *Water Res.* 46 (2012) 5813–33. doi:10.1016/j.watres.2012.08.002.

[123] C. Baronti, R. Curini, G. D'Ascenzo, A. Di Corcia, A. Gentili, R. Samperi, Monitoring Natural and Synthetic Estrogens at Activated Sludge Sewage Treatment Plants and in a Receiving River Water, *Environ. Sci. Technol.* 34 (2000) 5059–5066. doi:10.1021/es001359q.

[124] S. K. Behera, H. W. Kim, J.-E. Oh, H.-S. Park, Occurrence and removal of antibiotics, hormones and several other pharmaceuticals in wastewater treatment plants of the largest industrial city of Korea, *Sci. Total Environ.* 409 (2011) 4351–4360. doi:10.1016/j.scitotenv.2011.07.015.

[125] A. D. Vethaak, J. Lahr, S. M. Schrap, A. C. Belfroid, G. B. J. Rijs, A. Gerritsen, et al., An integrated assessment of estrogenic contamination and biological effects in the aquatic environment of The Netherlands, *Chemosphere.* 59 (2005) 511–524. doi:10.1016/j.chemosphere.2004.12.053.

[126] R. Céspedes, S. Lacorte, A. Ginebreda, D. Barceló, Occurrence and fate of alkylphenols and alkylphenol ethoxylates in sewage treatment plants and impact on receiving waters along the Ter River (Catalonia, NE Spain), *Environ. Pollut.* 153 (2008) 384–392. doi:10.1016/j.envpol.2007.08.026.

[127] H. Chang, Y. Wan, S. Wu, Z. Fan, J. Hu, Occurrence of androgens and progestogens in wastewater treatment plants and receiving river waters: Comparison to estrogens, *Water Res.* 45 (2011) 732–740. doi:10.1016/j.watres.2010.08.046.

[128] M. Clara, N. Kreuzinger, B. Strenn, O. Gans, H. Kroiss, The solids retention time—a suitable design parameter to evaluate the capacity of wastewater treatment plants to remove micropollutants, *Water Res.* 39 (2005) 97–106. doi:10.1016/j.watres.2004.08.036.

[129] G. D'Ascenzo, A. Di Corcia, A. Gentili, R. Mancini, R. Mastropasqua, M. Nazzari, et al., Fate of natural estrogen conjugates in municipal sewage transport and treatment facilities, *Sci. Total Environ.* 302 (2003) 199–209. doi:10.1016/S0048-9697(02)00342-X.

[130] M. P. Fernandez, T.-N. Noguerol, S. Lacorte, I. Buchanan, B. Piña, Toxicity identification fractionation of environmental estrogens in waste water and sludge using gas and liquid chromatography coupled to mass spectrometry and recombinant yeast assay, *Anal. Bioanal. Chem.* 393 (2009) 957–968. doi:10.1007/s00216-008-2516-8.

[131] G. Gatidou, N. S. Thomaidis, A. S. Stasinakis, T. D. Lekkas, Simultaneous determination of the endocrine disrupting compounds nonylphenol, nonylphenol ethoxylates, triclosan and bisphenol A in wastewater and sewage sludge by gas chromatography–mass spectrometry, *J. Chromatogr. A.* 1138 (2007) 32–41. doi:10.1016/j.chroma.2006.10.037.

[132] M. J. Gómez, M. J. Martínez Bueno, S. Lacorte, A. R. Fernández-Alba, A. Agüera, Pilot survey monitoring pharmaceuticals and related compounds in a sewage treatment plant located on the Mediterranean coast, *Chemosphere.* 66 (2007) 993–1002. doi:10.1016/j.chemosphere.2006. 07.051.

[133] J. Heidler, R. U. Halden, Mass balance assessment of triclosan removal during conventional sewage treatment, *Chemosphere.* 66 (2007) 362–369. doi:10.1016/j.chemosphere. 2006. 04.066.

[134] Y. Horii, J.L. Reiner, B.G. Loganathan, K. Senthil Kumar, K. Sajwan, K. Kannan, Occurrence and fate of polycyclic musks in wastewater treatment plants in Kentucky and Georgia, USA, *Chemosphere.* 68 (2007) 2011–2020. doi:10.1016/j.chemosphere. 2007.04.054.

[135] A. O. Ifelebuegu, The fate and behavior of selected endocrine disrupting chemicals in full scale wastewater and sludge treatment unit processes, *Int. J. Environ. Sci. Technol.* 8 (2011) 245–254. doi:10.1007/BF03326213.

[136] T. Isobe, H. Nishiyama, A. Nakashima, H. Takada, Distribution and behavior of nonylphenol, octylphenol, and nonylphenol monoethoxylate in Tokyo metropolitan area: Their association with aquatic particles and

sedimentary distributions, *Environ. Sci. Technol.* 35 (2001) 1041–1049. doi:10.1021/es001250i.

[137] B. Kasprzyk-Hordern, R. M. Dinsdale, A. J. Guwy, The removal of pharmaceuticals, personal care products, endocrine disruptors and illicit drugs during wastewater treatment and its impact on the quality of receiving waters, *Water Res.* 43 (2009) 363–380. doi:10.1016/j.watres.2008.10.047.

[138] S. D. Kim, J. Cho, I. S. Kim, B. J. Vanderford, S. A. Snyder, Occurrence and removal of pharmaceuticals and endocrine disruptors in South Korean surface, drinking, and waste waters, *Water Res.* 41 (2007) 1013–1021. doi:10.1016/j.watres.2006.06.034.

[139] M. Kostich, R. Flick, J. Martinson, Comparing predicted estrogen concentrations with measurements in US waters, *Environ. Pollut.* 178 (2013) 271–277. doi:10.1016/j.envpol.2013.03.024.

[140] Z. Liu, Y. Kanjo, S. Mizutani, Removal mechanisms for endocrine disrupting compounds (EDCs) in wastewater treatment — physical means, biodegradation, and chemical advanced oxidation: A review, *Sci. Total Environ.* 407 (2009) 731–748. doi:10.1016/j.scitotenv. 2008.08.039.

[141] N. Nakada, T. Tanishima, H. Shinohara, K. Kiri, H. Takada, Pharmaceutical chemicals and endocrine disrupters in municipal wastewater in Tokyo and their removal during activated sludge treatment, *Water Res.* 40 (2006) 3297–3303. doi:10.1016/j.watres.2006.06.039.

[142] Y. Nie, Z. Qiang, H. Zhang, W. Ben, Fate and seasonal variation of endocrine-disrupting chemicals in a sewage treatment plant with A/A/O process, *Sep. Purif. Technol.* 84 (2012) 9–15. doi:10.1016/j.seppur.2011.01.030.

[143] G.-G. Ying, R. S. Kookana, Triclosan in wastewaters and biosolids from Australian wastewater treatment plants, *Environ. Int.* 33 (2007) 199–205. doi:10.1016/j.envint.2006.09.008.

[144] C.-P. Yu, K.-H. Chu, Occurrence of pharmaceuticals and personal care products along the West Prong Little Pigeon River in east Tennessee, USA, *Chemosphere.* 75 (2009) 1281–1286. doi:10.1016/j.chemosphere.2009.03.043.

[145] B. Jarošová, L. Bláha, J. P. Giesy, K. Hilscherová, What level of estrogenic activity determined by in vitro assays in municipal waste waters can be considered as safe?, *Environ. Int.* 64 (2014) 98–109. doi:10.1016/j.envint.2013.12.009.

[146] B. Blair, J. Kehl, R. Klaper, Assessing emerging wastewater regulations to minimize the risk from pharmaceuticals and personal care products, *Manag. Environ. Qual. An Int. J.* 26 (2015) 966–983. doi:10.1108/MEQ-12-2014-0171.

[147] I. Bhati, N. G. Dhawan, R. K. Maheshwari, Greener Route to Prevent Pharmaceutical Pollution, *Int. J. Pharm. Chem. Sci.* 2 (2013) 1781–1787. http://www.ijpcsonline.com/files/23-674.pdf.

[148] B. Dunn, S. Miller, F. de Leon, Great Lakes-St. Lawrence River Basin Roadmap on Toxic Chemicals: Advancing Prevention by Promoting Safer Alternatives, 2015.

[149] N. El-Khateeb, The Condition of Streams and Prospects for Restoration in Palestine, in: A. Tal, A. Abed Rabbo (Eds.), *Water Wisdom Prep. Groundwork Coop. Sustain. Water Manag.* Middle East, New Brunswick, NJ and London, England, 2010.

[150] L. Teclaff, Evolution of the river basin concept in national and international water law, *Nat. Resour. J.* 36 (1996).

[151] F. Molle, P. Wester, P. Hirsch, River Basin Development and Management, in: D. Molden (Ed.), *Water Food, Water Life A Compr. Assess. Water Manag. Agric.,* International Water Management Institute, 2006.

[152] WHO-- World Health Organization, WHO Guidelines for the Safe Use of Wastewater, Excreta and Greywater, 2006.

[153] S. Amr, Health for All? Educational Video., Galilee Society, 1993.

[154] K. Tubail, Sewage Treatment, in: A. Tal, A. Alfred Abed- Rabbo (Eds.), Water Wisdom, a New Menu *Palest. Isr. Coop. Water Manag.,* Rutgers University Press, New Brunswick, NJ, 2010: pp. 216–228.

[155] S. Afifi, Wastewater reuse in agriculture in North Gaza and environment, *Egypt J. Soil Sci.* 40 (2000) 132–148.

[156] A. E. Al-Juaidi, D. E. Rosenberg, J. J. Kaluarachchi, Water management with wastewater treatment and reuse, desalination, and conveyance to counteract future water shortages in the Gaza Strip, *Int. J. Water Resour. Environ. Eng.* 3 (2011) 266–282.

[157] A. Yasin, Status of Wastewater and Challenges in Palestine: Westbank, (2012).

[158] A. Jayousi, The Oslo II Accords in Retrospect: Implementation of the Water Provisions in the Israeli and Palestinian Interim Peace Agreements, in: A. Tal, A. Abed Rabbo (Eds.), *Water Wisdom Prep. Groundwork Coop. Sustain. Water Manag.* Middle East, Rutgers University Press, New Brunswick, NJ and London, England, 2010.

[159] D. Kerret, An Israeli Retrospective, in: A. Tal, A. Abed Rabbo (Eds.), *Water Wisdom Prep. Groundwork Coop. Sustain. Water Manag.* Middle East, Rutgers University Press, New Brunswick, NJ and London, England, 2010.

[160] I. Fischhendler, S. Dinar, D. Katz, The Politics of Unilateral Environmentalism : Cooperation and Conflict over Water Management along the Israeli-Palestinian Border, *Glob. Environ. Polit.* 11 (2011) 36–61. http://www.mitpressjournals.org/doi/abs/10.1162/GLEP_a_00042.

[161] I. Fischhendler, Legal and institutional adaptation to climate uncertainty: a study of international rivers, *Water Policy.* 6 (2004) 281–302. http://test.iwaponline.com/wp/00604/0281/006040281.pdf.

[162] A. Schalimtzek, I. Fischhendler, Dividing the cost burden of environmental services: The Israeli–Palestinian wastewater regime, *Env. Polit.* 18 (2009) 612–632. doi:10.1080/09644010903007468.

[163] N. Ghbn, The Coming Age of Desalination for Gaza: Visions, Illusions, and Reality, in: A. Tal, A.A. Rabbo (Eds.), *Water Wisdom Prep. Groundwork Coop. Sustain. Water Manag.* Middle East2, Rutgers University Press, New Brunswick, NJ and London, England, 2010.

[164] E. Feitelson, The Upcoming Challenge: Transboundary Management of the Hydraulic Cycle, Water. *Air. Soil Pollut.* 123 (2000) 533–549. doi:10.1023/A:1005269001462.

[165] S.P. Mumme, K. Collins, The La Paz Agreement 30 Years On, J. Environ. Dev. 23 (2014) 303–330. doi:10.1177/1070496514528801.

[166] J. a. Yoder, Bridging the European Union and Eastern Europe: Cross-border Cooperation and the Euroregions, *Reg. Fed. Stud.* 13 (2003) 90–106. doi:10.1080/13597560308559436.

[167] D. Jensen, S. Lonergan, eds., Assessing and Restoring Natural Resources In *Post-Conflict Peacebuilding,* 1st ed., Earthscan, New York, NY, 2012.

[168] United Nations Environment Programme, From Conflict to Peacebuilding: The role of natural resources and the environment, 2009. http://www.unep.org/pdf/pcdmb_policy_01.pdf.

Appendix A. EDC Concentrations According to Location in ng/L (Nanograms/Liter)

Location	Estrone	Estriol	Triclosan	4-tert- Octylphenol	BPA
Israel	0.3	0.03	5		5
Israel	0.5646	7.802	5		5
Israel	4.1	0.03	60.33333333		5
Israel	1.2168	2.56	28.66666667		2.333333333
Israel	1.426366667	3.326666667	74		5.3
Israel	0.8845	0.03	101		5
Israel	1.336733333	2.16	27.66666667		4.133333333
Israel	0.8026	0.03	5		5
Israel	1.85525	3.325	5		3.8
Israel	0.65	0.043333333	179.6666667	2.333333333	2.333333333
Israel	0.036666667	0.043333333	103.3333333	2.333333333	2.333333333
Israel	5.5192	10.4086	99.56666667		1
Israel	11.8901	4.0587	70.16666667		17.33333333
Israel	5.816666667	0.525	5		17.33333333
Israel	3.366666667	0.293333333	397	2.333333333	8.666666667
Israel	0.036666667	0.043333333	150.3333333	2.333333333	2.333333333
Israel	3.65	0.043333333	170	2.333333333	307.3333333
Israel	2.32	0.043333333	136.6666667	2.333333333	2.333333333
Israel	4.2	5.843333333	103.3333333	2.333333333	2.333333333
Israel	4.233333333	0.043333333	5	2.333333333	2.333333333

Location	Estrone	Estriol	Triclosan	4-tert- Octylphenol	BPA
Israel	21.3	7.733333333	784	2.333333333	2.333333333
Israel	0.05	0.05	5	1	1
Israel	8.3	4.993333333	368	2.333333333	2.333333333
Israel	4.903333333	1.393333333	276.3333333	2.333333333	2.333333333
Israel	1.6	4.6	5	8.666666667	2.333333333
Israel	4.866666667	2.293333333	201.6666667	2.333333333	2.333333333
Israel	1.6	3.26	136.6666667	2.333333333	5.333333333
Israel	0.556533333	0.043333333	5		2.333333333
Israel	1.931333333	0.043333333	5		2.333333333
Israel	2.253333333	1.76	2.543	2.333333333	2.333333333
Israel	5.803333333	1.86	2.55	2.333333333	1
Israel	1.725	0.05	5	1	1
PA West Bank	4.948443953	2.10463817	77.33333333	3	3
PA West Bank	4.754094374	0.04	5	3	3
PA West Bank	3.04013461	0.03	174	5	5
PA West Bank	1.116441284	29.2	5	1	1
PA West Bank	0.01	0.03	5	5	5
PA West Bank	0.01	0.03	5	5	5
PA West Bank	0.01	0.03	5	5	5
PA West Bank	0.01	0.03	5	5	5
PA West Bank	0.01	0.03	5	5	5
Average	2.986275	2.429407938	93.04454472	3.061728395	11.46910569

Appendix B. EDC Concentrations According to Location Expressed as EEQs (ng/L EEQ)

Location	Estrone	Estriol	Triclosan	4-tert- Octylphenol	BPA	EEQ Total per location
Israel	0.033	0.0033	0.00000025		0.00195	0.03825025
Israel	0.062106	0.85822	0.00000025		0.00195	0.92227625
Israel	0.451	0.0033	3.01667E-06		0.00195	0.456253017
Israel	0.133848	0.2816	1.43333E-06		0.00091	0.416359433
Israel	0.156900333	0.365933333	0.0000037		0.002067	0.524904367
Israel	0.097295	0.0033	0.00000505		0.00195	0.10255005
Israel	0.147040667	0.2376	1.38333E-06		0.001612	0.38625405
Israel	0.088286	0.0033	0.00000025		0.00195	0.09353625
Israel	0.2040775	0.36575	0.00000025		0.001482	0.57130975
Israel	0.0715	0.004766667	8.98333E-06	0.000979133	0.00091	0.078164783
Israel	0.004033333	0.004766667	5.16667E-06	0.000979133	0.00091	0.0106943
Israel	0.607112	1.144946	4.97833E-06		0.00039	1.752452978
Israel	1.307911	0.446457	3.50833E-06		0.00676	1.761131508
Israel	0.639833333	0.05775	0.00000025		0.00676	0.704343583
Israel	0.370333333	0.032266667	0.00001985	0.000979133	0.00338	0.406978983
Israel	0.004033333	0.004766667	7.51667E-06	0.000979133	0.00091	0.01069665
Israel	0.4015	0.004766667	0.0000085	0.000979133	0.11986	0.5271143
Israel	0.2552	0.004766667	6.83333E-06	0.000979133	0.00091	0.261862633
Israel	0.462	0.642766667	5.16667E-06	0.000979133	0.00091	1.106660967

Location	Estrone	Estriol	Triclosan	4-tert- Octylphenol	BPA	EEQ Total per location
Israel	0.465666667	0.004766667	0.00000025	0.000979133	0.00091	0.472322717
Israel	2.343	0.850666667	0.0000392	0.000979133	0.00091	3.195595
Israel	0.0055	0.0055	0.00000025	0.000419629	0.00039	0.011809879
Israel	0.913	0.549266667	0.0000184	0.000979133	0.00091	1.4641742
Israel	0.539366667	0.153266667	1.38167E-05	0.000979133	0.00091	0.694536283
Israel	0.176	0.506	0.00000025	0.003636781	0.00091	0.686547031
Israel	0.535333333	0.252266667	1.00833E-05	0.000979133	0.00091	0.789499217
Israel	0.176	0.3586	6.83333E-06	0.000979133	0.00208	0.537665967
Israel	0.061218667	0.004766667	0.00000025		0.00091	0.066895583
Israel	0.212446667	0.004766667	0.00000025		0.00091	0.218123583
Israel	0.247866667	0.1936	1.2715E-07	0.000979133	0.00091	0.443355927
Israel	0.638366667	0.2046	1.275E-07	0.000979133	0.00039	0.844335928
Israel	0.18975	0.0055	0.00000025	0.000419629	0.00039	0.196059879
PA West Bank	0.544328835	0.231510199	3.86667E-06	0.001258886	0.00117	0.778271786
PA West Bank	0.522950381	0.0044	0.00000025	0.001258886	0.00117	0.529779517
PA West Bank	0.334414807	0.0033	0.0000087	0.002098143	0.00195	0.34177165
PA West Bank	0.122808541	3.212	0.00000025	0.000419629	0.00039	3.33561842
PA West Bank	0.0011	0.0033	0.00000025	0.002098143	0.00195	0.008448393
PA West Bank	0.0011	0.0033	0.00000025	0.002098143	0.00195	0.008448393
PA West Bank	0.0011	0.0033	0.00000025	0.002098143	0.00195	0.008448393
PA West Bank	0.0011	0.0033	0.00000025	0.002098143	0.00195	0.008448393
PA West Bank	0.0011	0.0033	0.00000025	0.002098143	0.00195	0.008448393
Average	0.330012871	0.269063411	4.65223E-06	0.001284789	0.004472951	0.604838675

BIOGRAPHICAL SKETCH

Name: Vivian Futran Fuhrman
Affiliation: PhD Candidate, Ben Gurion University

SUMMARY: Diverse experience driving environmental projects, research, and events at the intersection of science and policy. Expertise in human health risk assessment and exposure to endocrine disrupting chemicals. Familiarity with academic, private and government perspectives on water scarcity, public health and energy efficiency in buildings.

Education:

Ben-Gurion University, Beersheva, Israel PhD, Environmental Risk and Policy	2011-2016 (expected)
University of Pennsylvania, Philadelphia, PA MES (Master of Environmental Studies)	2010-2011
Arava Institute for Environmental Studies, Kibbutz Ketura, Israel Certificate in Environmental Studies	2009
Duke University, Durham, NC	2003-2007

BA, Political Science/International Relations

Address: 205 Walden Street, Apt. 4k, Cambridge, Massachusetts, 02140, USA

Research Experience:

Ben-Gurion University, Israel 2011-2015
PhD Candidate

- Dissertation topics: How endocrine disrupting chemicals (EDCs) challenge traditional risk assessment; Modeling relative exposure to estrogenic EDCs from recycled wastewater for human health risk assessment; Regional and global policy implications of EDC exposure from wastewater recycling and cost efficiency analysis of treatment options

- Drylands, Deserts & Desertification Conference: Co-convened session on Zero Net Land Degradation for UN Convention to Combat Desertification, co-authored conference session outcome (see Publications)
- Green Building Project: Year-long government funded research on global green building initiatives and incentive

Book Project, Arlington, VA 2008-2009
Author's Assistant

- Worked with Dr. Jordan D. Lewis, global alliance expert, on multidisciplinary book addressing human group relations

Duke University Psychology Laboratories, Durham, NC 2006
Lab Assistant

- Infant language acquisition project

Eagleton Institute of Political Science at Rutgers University
New Brunswick, NJ 2005

Research Intern

- Coordinated primary research and data collection on Young Elected Leaders Project

Professional Experience:

Product Stewardship Institute, Boston, MA
2015-present
Associate for Policy and Programs

Global Water Alliance, Philadelphia, PA 2010-2011
Program Manager

EPA Region 3 Office of Water, Philadelphia, PA 2010-2011
Student Intern

ManageEnergy, Arglington, VA 2008- 2009

Special Projects Manager

ICF International (Environmental Consulting firm),
Fairfax, VA 2007-2008
Research Assistant

Honors:

- Ben Gurion University Negev Scholarship, Faran Fellow
- University of Pennsylvania, Masters of Environmental Studies Program, Award: Outstanding Graduate in Master of Environmental Studies Program

Publications Last Three Years:

Futran Fuhrman, V., “Endocrine Disrupting Chemicals in Treated Wastewater: Risk Assessment and Management Implications in Israel and the West Bank,” Ben Gurion University, PhD Dissertation (2015).

Futran Fuhrman, V., A. Tal, S. Arnon, “Why Endocrine Disrupting Chemicals (EDCs) Challenge Traditional Risk Assessment and How to Respond,” Journal of Hazardous Materials 286 (2015): 589-611.

Chasek, P., U. Safriel, S. Shikongo, V. Futran Fuhrman, “Operationalizing Zero Net Land Degradation: The Next Stage in International Efforts to Combat Desertification?” Journal of Arid Environments 112 (2015): 5-13.

Futran Fuhrman, V., “Why we need to act on desertification today,” British Council, VOICES online blog (2015).

Futran Fuhrman, V., “Why Endocrine Disrupting Chemicals Challenge Traditional Risk Assessment and How to Respond,” Poster presentation, ISEE-Europe Conference on Environmental Epidemiology, Barcelona, Spain (2014).

Futran, V., “Tackling Water Scarcity: Israel's Wastewater Recycling as a Model for the World's Arid Lands?” Australia Water Engineers magazine June/July (2013).

Futran, V., “Tackling Water Scarcity: Israel's Wastewater Recycling as a Model for the World's Arid Lands?” UNESCO's Global Water Forum (2013).

In: Endocrine Disrupting Chemicals
Editor: Sabrina Johnston
ISBN: 978-1-63485-231-9

Chapter 3

EFFECT OF ENDOCRINE DISRUPTING CHEMICALS ON PREGNANCY

***Tandra R. Chakraborty*[1], *Sanjoy Chakraborty*[2] *and Natalia Klimova*[1]**

[1]Department of Biology, Adelphi University, One South Avenue, Garden City, New York, NY US

[2]Department of Biological Sciences, New York City College of Technology/CUNY, New York, NY, US

ABSTRACT

Endocrine disrupting chemicals (EDC's) are exogenous chemicals that are either plant-derived or man-made, and can alter the functions of the endocrine system and cause various health problems by interfering with the synthesis, metabolism, binding, or cellular responses similar to natural estrogens. Industrialization has introduced a battery of EDC's to the environment which has shown to have drastic outcomes on human health particularly reproduction and pregnancy. Successful pregnancy requires proper orchestration and adaptations of the endocrine system. This article summarizes the role of the EDC's [polybrominated diphenyl ethers (PBDE), 2, 3, 7, 8-tetrachlorodibenzo-p-dioxin (TCDD), bisphenol A (BPA), phthalates (DEHP)s] on pregnancy and birth outcomes, included but not limited to fertility problems, premature birth, low birth weight, and miscarriage along with *in-vitro*, animal, and epidemiological studies. Phytoestrogens are included in the discussion to compare effects of naturally occurring endocrine disrupting chemicals to the man-made ones. Several research demonstrates that fetal exposure to EDC's may negatively affect reproductive health for the lifetime of the organism, increasing chances of reduced fertility, negative pregnancy outcomes, and poor birth outcomes. Epidemiological experiments often associate high concentration of the chemicals with elevated chances of adverse outcomes and reduced fertility. However, there are many missing and conflicting information in this area. Despite uncertainties and knowledge gaps, the available evidence suggests that limiting exposure to EDC's especially during pregnancy is advisable since it can show its affect even two generations latter.

INTRODUCTION

Industrialization introduced technological and economical advances that allowed many countries to develop rapidly improving quality of life of many people; however this process caused increased levels of pollution of our environment. Some of the industrial byproducts and pollutants which are artificially made contaminate the environment and also have negative impact on human health and wildlife affecting the endocrine system in particular. The most common ways of exposure to many toxic agents are through inhalation of contaminated air, ingestion of contaminated food or water, or direct contact.

In this paper, the focus will be on polybrominated diphenyl ethers (PBDE), 2, 3, 7, 8-tetrachlorodibenzo-p-dioxin (TCDD), bisphenol A (BPA), phthalates (DEHP)s and phytoestrogens all of which are endocrine disruptors. Despite of advances in medicine and technology, infertility and adverse pregnancy outcomes remain an ongoing problem. Chronic exposure to low levels of endocrine disrupting chemicals (EDC's) could be one of the reasons leading to fertility problems, pregnancy complications and preterm birth. These complications in turn may lead to poor birth outcomes including neonates with low birth weight and neonates that are small for gestation. In addition, fetuses may be more sensitive for the effects of the exogenous compounds because of organs development (organogenesis) and EDC's may disturb normal developing processes. The negative effects may last into adult life, affecting reproductive and other systems. In addition, the effects can even last even two generations latter. Therefore, it is important to evaluate if chronic exposure to environmental pollutants may negatively affect pregnancy.

INFERTILITY, PREGNANCY COMPLICATIONS, AND POOR PREGNANCY OUTCOMES

Infertility is inability to conceive after one year of sexual activity without the use of any contraception. According to Centers of Disease Control and Prevention, about 6% of married women of reproductive age in the United States are unable to get pregnant after one year of trying (CDC, 2014). The most common reasons of fertility problems are diminished function of ovaries in females and abnormal sperm in males. It has been demonstrated that certain EDC's may disturb early oogenesis and follicle formation in female fetuses and causing sperm morphology alterations in males (Bruner-Tran et al., 2014; Hunt et al., 2012).

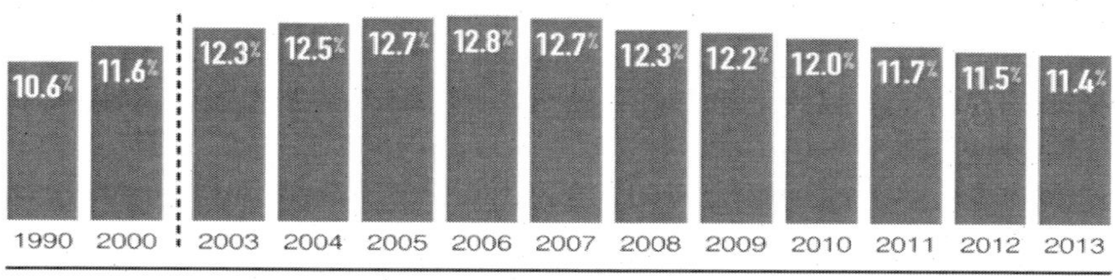

Figure 1. Pre-term rates in the United States 1990, 2000, 2003-2013. Source National Center for Health Statistics, 1990-2013 final natality data. Adapted from March of Dimes (2015).

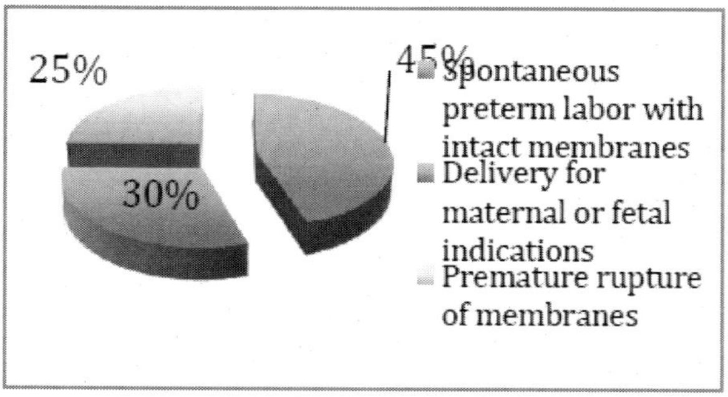

Figure 2. Factors associated with pre-term birth. Adapted from Goldenberg et al., (2008).

Some of the most prevalent pregnancy complications are miscarriage, placental abruptions, preeclampsia, and preterm-labor. Miscarriage is defined as spontaneous loss of a fetus at less than 20 weeks of pregnancy. In many cases miscarriage is due to genetic factors, followed by endocrine problems, anatomic and immunologic abnormalities, infection, or major disease (Roberts et al., 2000). Placental abruption is condition in which placenta prematurely separates from the uterine wall. It can cause serious bleeding and deficit of oxygen and nutrients to the fetus. According to U.S. National Library of Medicine, preeclampsia is a condition in which pregnant woman develops high blood pressure (higher than 140/90) and protein in her urine after gestation day 20 (MedlinePlus, 2014). In most cases preeclampsia goes away after delivery, suggesting that abnormalities in placenta are the major contributors to the syndrome. Possibly, EDC's may contribute to abnormal placentation.

Preterm birth is childbirth that occurs at less than 37 completed weeks of gestation (World Health Organization [WHO], 2014). Preterm birth is a major factor of neonatal mortality and morbidity and has long-term adverse consequences for health. About 1 million neonates die each year due to complications associated with preterm birth (WHO, 2014). According to National Center for Health Statistics, pre-term rates are very high in the United States and since 1980s they have increased by about 25%. In 2012, 11.5% (1 in 9 babies) of all births were delivered preterm (Figure 1) (March of Dimes, 2014). Even though the rate has been declining since its peak in 2006, which was at 12.8%, it is still one of the highest in the developed nations.

There are several major causes that lead for pre-term birth such as maternal or fetal medical indications which accounts for about 30% of all pre-term births cases, spontaneous pre-term labor with intact membranes which accounts for 45% cases of pre-term birth, and 25% cases due to premature rupture of membranes (Figure 2) (as cited in Goldenberg et al., 2008).

The most common conditions that require medically indicated pre-term births are preeclampsia, fetal distress, small-for-gestational age fetus, and placental abruptions (Avanth & Vintzileos, 2006). According to WHO, low birth weight (LBW) is weight at birth of less than 2,500 grams and small-for-gestational age (SGA) fetus is a fetus with weight below the 10th percentile for the gestational age (WHO, 2014). According to Centers for Disease Control and Prevention (CDC), 8% of babies were born low birth weight in 2013 in the U.S. (2014). Two major reasons of infants' low birth weight are prematurity and uterine growth restrictions. Premature, SGA, and LBW infants are at greater risk for developing medical complications and have a higher risk for death (Regev et al., 2004). In many cases causes of spontaneous preterm birth and premature rupture of membranes are unknown; however, severe inflammatory response caused by intrauterine infections is thought to be one of the main precursors leading to it. Inflammation also has been linked with uterine growth retardation and SGA fetus.

INFLAMMATION AND ROLE OF PRO-INFLAMMATORY MOLECULES IN PREGNANCY COMPLICATIONS

Successful pregnancy requires adaptation of immune system to allow fetus to grow. There is a delicate balance established that allows the immune system to protect woman's body from infections and at the same prohibits it from rejecting the fetus. External stressors may disturb the balance. For example, during bacterial infection, production of pro-inflammatory molecules is increased causing inflammation at the maternal-fetal interface. A balance between pro-inflammatory molecules such as interleukin -1b (IL-1b), tumor necrosis factor alfa (TNF-α), interleukin 6 (IL-6), and interleukin 8 (IL-8) and anti-inflammatory mediators such as interleukin -10 (IL-10) is important in maintaining healthy pregnancy. Previously, it was described that deregulation of cytokines may increase the risk of pre-term birth (Pomini et al., 1999). Also, chronic low-grade inflammation is linked to development of preeclampsia and uterine growth restriction. Pregnant women who had higher levels of c-reactive protein, one of the markers of inflammation, in their blood during first trimester had significantly higher chances of developing preeclampsia and uterine growth restriction (Tjoa et al., 2003). Some of the EDC's have been found to be able to increase production of inflammatory markers or modify immune response in various types of cells. For instance, a well-known for its adverse health effects insecticide DDT has been shown to induce production of IL-1b, IL-6, and TNF-a cytokines in macrophages in a dose dependent manner (Kim et al., 2004). Zhang et al., 2014 demonstrated that one mechanism of TCDD neurotoxicity is by induction of pro-inflammatory environment in astrocytes. Similarly, environmental pollutants may be able to induce inflammation in the uterus, which may increase the risk of pregnancy complications. Peltier et al., 2012 proposed that possible mechanism of TCDD and PBDE toxicity in placenta that causes pregnancy complications is by decreasing production of IL-10. Reduced levels of IL-10 have been shown to contribute to abnormal placentation and vascularization, which are one of the key precursors of preeclampsia (Kalkunte et al., 2011).

Oxidative Stress and Pregnancy Complications

Oxidative stress occurs when the ratio of oxidants and antioxidants is not balanced. Excessive production of reactive oxygen species (ROS) or reduced levels of antioxidants can shift the balance to create oxidative stress. Normal physiological processes such as electron transfer reactions in mitochondria generate ROS. Enzymes superoxide dismutase (SOD), catalase, and glutathione peroxidase are important antioxidants. There are many factors that can shift the balance and contribute to generation of oxidative damage, including poor nutrition, obesity, drugs, smoking, alcohol, and environmental pollutants (Figure 3). When tissues are damaged by oxidative stress they become necrotic. Necrosis induces inflammatory response in the surrounding cells and tissues. Thus, oxidative stress is always accompanied by inflammatory reactions and the same key inflammatory mediators are involved such as IL-1β, TNF-α, IL-8, and IL-6 (Menon, 2012). As it was discussed above excessive production of pro-inflammatory cytokines may increase risks of pre-term labor, preeclampsia, and SGA fetus. Therefore, mechanism causing pregnancy complications due to oxidative damage may involve similar pathways to the one caused by bacterial infection. In addition, women who delivered prematurely had lower levels of antioxidant enzymes in comparison to women who delivered at term (Menon, 2012). Presence of oxidative stress in placenta has been linked to preeclampsia and uterine growth restriction (Burton et al., 2009). Also, many studies have shown that ROS negatively affect male and female fertility. ROS may reduce sperm development, maturation, and quality in males and cause unexplained infertility in females (as cited in Agarwal et al., 2012).

Hormones and Their Role in Infertility, Adverse Pregnancy and Outcomes

Hormones play a critical role in pregnancy. Successful pregnancy also requires changes and adaptations in the endocrine system. Placenta is a temporary endocrine organ that is responsible for implantation of the embryo to the uterus and production of hormones that support pregnancy during early gestation, and it provides nutrient and oxygen transport to the developing fetus from mother during later gestations (Myren et al., 2007). It produces hormones estrogen, progesterone, and chorionic gonadotrophin (hCG), also factors such as cytokines, chemokines, eicosanoids, pregnancy-associated proteins, and many others. Maintaining proper levels of progesterone and estrogen during pregnancy is very important for the positive outcomes. About 8 to 12% of all pregnancy losses are due to endocrine alterations (Arredondo & Noble, 2006). Progesterone plays an important role in continuation of pregnancy by inhibiting uterine contractions. Ovaries produce progesterone during the first trimester of pregnancy; but by the 9th week of pregnancy placenta is the main source of the hormone. Sharp decline of progesterone levels may trigger labor. Estrogen is also important in maintaining uterus, cervix, and placenta during pregnancy. Human chorionic gonadotrophin (hCG) is produced by trophoblast cells. It is created in high amounts in early pregnancy. Because of the ability to accumulate in placenta, EDC's may interfere with normal placental functions by deregulating pregnancy hormone levels. For example, TCDD increases estradiol secretion in placental tissue cultures (Augustowska et al., 2003). Therefore, analyzing levels of the toxins in placenta may demonstrate the exposure of the mother and fetus to these chemicals throughout the duration of pregnancy. Endogenous hormonal levels need to be tightly regulated during pregnancy especially during the period of development of sensitive fetal organs. Exposure to low doses of exogenous estrogens may negatively affect fetal ovary development, which in turn may cause fertility problems later in life. Rajapakse et al., 2002 suggested that exposure to multiple estrogenic compounds may have an additive effect in the organism. The authors examined effect of 11 estrogenic compounds including BPA and phytoestrogen genisin on function of 17β-estradiol using yeast reporter gene assay. The estrogenic compounds at low doses enhanced the hormone action. This evidence implies that exposure to even very low doses of estrogenic chemicals may be dangerous for reproductive health because their effects may add up with other endogenous or exogenous estrogens and alter normal function of hormones. In addition, it has been shown that some hormonal imbalances adversely affect fertility. For example, inadequate levels of progesterone may cause infertility problems in women (Roberts & Murphy, 2000).

Thyroid hormones have direct effect on ovaries and interact with sex hormones. Thyroid hormones play an important part in normal pregnancy, and their deregulation can lead to infertility and increased risk of early miscarriage. Also, hypothyroidism has been linked with increased morbidity and a poorer outcome of pregnancy, and hyperthyroidism has been connected with pre-term birth, placental abruption, fetal growth restriction, and stillbirth (Arredondo et al., 2006; Poppe et al., 2008). High levels of PBDEs in blood serum have been associated with thyroid hormones alterations (Xu et al., 2014). Treating human cell lines with environmentally relevant doses of BPA demonstrated that the chemical has anti-thyroid hormone effect; BPA displaced thyroid hormone T3 from its receptor (as cited in Welshons et al., 2006). Therefore, endocrine-modulating chemicals may cause complication by disturbing normal balance of thyroid and sex hormones.

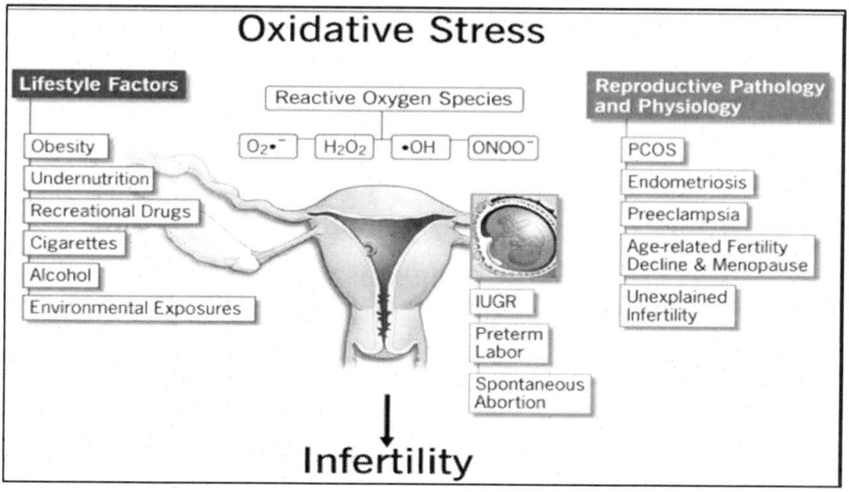

Figure 3. Summary of effects of oxidative stress on female reproductive system. Adapted from Agarwal, et al., (2012).

In this paper, the role of the EDC's in pregnancy and birth outcomes, including and not limited to fertility problems, premature birth, low birth weight, and miscarriage will be evaluated. Phytoestrogens are included in the discussion to compare effects of naturally occurring endocrine disrupting chemicals to the man-made ones. How each toxin affects pregnancy will be discussed in detail. Epidemiological, animal, and *in vitro* studies will be used.

EDC's and Its Effect on Pregnancy and Birth Outcome

Endocrine disruptors are exogenous chemicals that are usually man-made, that are present in various pesticides, metals, additives or contaminants in food, and personal care products. They can alter the functions of the endocrine system and cause various health problems by interfering with the synthesis, metabolism, binding, or cellular responses similar to natural estrogens. Previous studies have shown that EDC's have the ability to interfere with female and/or male reproductive system due to their endocrine disrupting properties (Caserta et al., 2008), abnormal growth patterns and neurodevelopmental delays in children, as well as changes in immune function. Some of the major EDC's affecting pregnancies are discussed below.

A. Polybrominated Diphenyl Ethers (PBDEs)

PBDE are effective flame-retardants that have been used in a large variety of consumer products since 1960s. The compound is composed of two phenyl rings linked by an oxygen bridge (Figure 4). PBDE are made as penta-, octo-, and deca-brominated mixtures. Pentabromodiphenyl ether mixture (penta-BDE) contains PBDEs that have four or five bromine atoms and it is used in epoxy resins, polyesters, flexible polyurethane foams in mattresses and pillows, and textiles. Octa-BDE is used in thermosets and polycarbonate. Deca-BDE is used to prevent fires in plastics for electrical and electronic equipment and in textiles. These chemicals are widely used by many manufacturing companies due to their fire preventive properties. If plastic products containing PBDE are exposed to excessive heat, the bromine atoms detach from the rings and this helps to prevent fire from spreading, giving time for an escape (Alaee et al., 2003). In 1999 in the European Union, annual consumption of penta-BDE was 150 metric tons, octa-BDE was 400 metric tons, and deca-BDE was 7000 metric tons (Bromine Science and Environmental Forum [BSEF], 2000). Because PBDE are not covalently bound to plastic products, overtime the compounds come off and become airborne. Humans are exposed to the pollutant through dust and by consumption of contaminated food. PBDEs are very lipophilic and persistent in environment, and accumulate in soil and organisms. In 2004, some companies voluntarily phased out use of the penta-BDE and octa-BDE due to negative health effects. Penta-BDE mixtures are believed to be the most toxic (Environment Protection Agency [EPA], 2014). Previous studies have shown that the chemical disrupts development of nervous system and also acts as an endocrine disruptor (Kuriyama et al., 2005). Also, because of its similarity to thyroid hormone, it affects thyroxine (T4) metabolism. Rats exposed to pentabromodiphenyl ether mixture in utero had significantly lower levels of T4 (Kodavanti et al., 2010). Polychlorinated biphenyls (PCB) are very similar in structure to PBDE, the only difference is that PCB don't have the oxygen bride (Figure 4). Chlorine and bromine atoms both belong to halogen group in the periodic table and have similar properties. Research performed decades ago demonstrated that PCB is toxic to human health and the chemical was banned in 1979 (EPA, 2013). Because of the structural similarity, the chemicals may have comparable effects on health.

According to Agency for Toxic Substances and Disease Registry (ATSDR), the minimal risk level of acute oral exposure to PBDEs is 30 ug/kg/day and of oral and inhaled intermediate exposure to the toxin is 6 ug/kg/day and 6 mg/m3 (ATSDR, 2014). The agency has not assessed safe chronic exposure levels to the chemical. PBDE is detectable in blood, cord blood, placenta, amniotic fluid, and breast milk. Evaluating 15 pregnant women from Ann Arbor, Michigan, Miller et al., (2012) found that congeners BDE-47 and BDE-99 were present in all amniotic fluid samples. The mean concentration of toxins in these samples was 3.8 ug/ml amniotic fluid. In one study, the pollutant was detectable in all 43 placentas that were obtained from three different US states. The median concentration of PBDE in those placentas was 0.33 ng/g wet weight and the most prevalent congeners were BDE-47, followed by BDE -153, -99, -100, and -209 (Nanes et al., 2013). In cord blood obtained from neonates from China, BDE-209 congener was present at higher concentrations with median levels equal to 2.54 ng/g lipid, followed by BDE -47, -28, -183, and -153 congeners (Wu, et al., 2010).

FEMALE/MALE FERTILITY

Animal studies have shown that exposure to PBDE can permanently modify male and female reproductive system. Rat males exposed to a single dose of BDE-99 in utero had lower spermatid and sperm counts; those males also exhibited decreased sperm production by about 30% in comparison to unexposed males (Kuriyama et al., 2005). Female rats treated with 700 ug/kg BDE-47 on gestation day 6 exhibited reduced number of growing follicles and reduced serum estradiol levels (Talsnees et al., 2008).

In humans Harley et al., (2010) associated high levels of PBDE in blood serum and increased time to get pregnant.

PREGNANCY OUTCOMES (MISCARRIAGE, PREECLAMPSIA)

Johnson et al., (2012) measured PBDE levels in ovarian follicular fluid samples obtained from 65 women undergoing *in-vitro* fertilization. The authors suggested that the presence of BDE-153 in follicular fluid might be linked to unsuccessful embryo implantation. Not much information is available that addresses this issue. More studies are needed to evaluate the effect of PBDE on pregnancy complications.

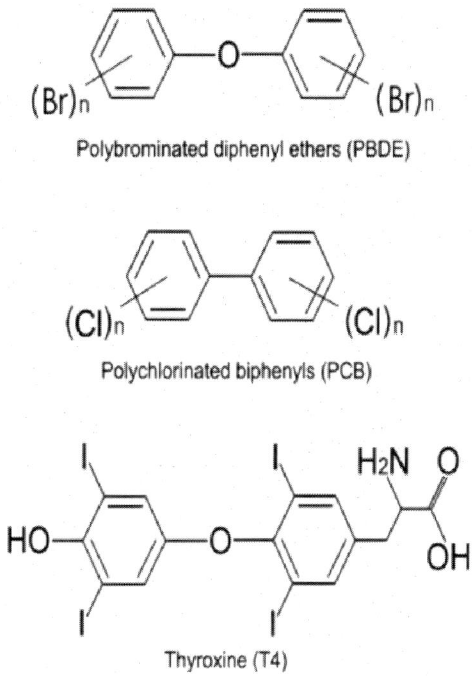

Figure 4. Similarity of PBDE structure to thyroid hormone T4.

BIRTH OUTCOMES (PRE-TERM BIRTH, LOW BIRTH WEIGHT, SGA)

Animal studies have shown that exposure to PBDE may reduce fetal birth weight. Pregnant rats exposed to 5 or 10 mg/kg body weight BDE-209 congener for 21 days from gestation day 0 delivered offspring rats that were significantly lower in weight in comparison to control animals (Du et al., 2015).

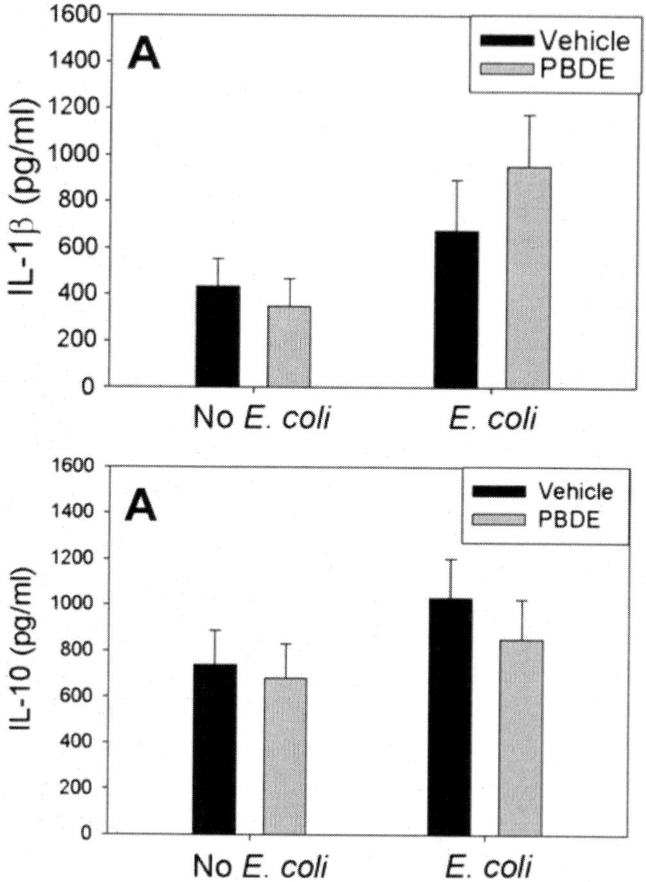

Figure 5. Effect of PBDE on IL-1b, IL-10 production in placental explants. Adapted from Peltier et al., (2012).

High concentrations of PBDE and high prevalence of BDE-28, -47, -99, -153, and -183 congeners in cord blood were correlated with significant increase in number of adverse birth outcomes such as premature delivery, low birth weight, and stillbirth (Wu et al., 2010). In the study, the median concentration of all PBDE detected in cord blood samples obtained from normal births was 9.882 ng/g lipid and the median concentration of PBDEs found in cord blood samples obtained from pregnancies with adverse outcomes was 41.97 ng/g lipid. Similarly, high levels of congener BDE-47 (323 pg/ml and higher) in blood plasma samples obtained from pregnant women significantly increased the odds of preterm birth (Peltier et al., 2014). Another study also correlated presence of BDE-47, -99, -100, and -153 congeners in breast milk obtained 3 weeks after delivery and reduced birth weight of neonates in Swedish women (Lignell et al., 2013).

As it was discussed above, intrauterine bacterial infection contributes to higher risks of pre-term delivery. It has been found that PBDEs may enhance inflammation at maternal fetal interface in the presence of infection (Peltier et al., 2012). The authors used second trimester placental explants to demonstrate that mixture of BDE-47, BDE-99, and BDE-100 congeners at final concentration of 2 uM in the presence of low dose E. *coli* caused dramatic increase of inflammatory cytokine IL-1b and reduction of anti-inflammatory cytokine IL-10 (Figure 5). This suggests that in the presence of this toxin, even low dose of bacteria may trigger exaggerated inflammatory response. Thus, PBDE may increase the risk of pre-term birth by augmenting inflammation response.

Recently, it has been shown that PBDEs by themselves can initiate inflammatory cascade in placental tissue. Exposure of first trimester trophoblast cells (HTR-8/SV neo) to BDE-47 congener for 4 hours significantly induced production of reactive oxygen species (ROS) and superoxide anions (Park et al., 2014). Also, in the same experiment exposure of trophoblast cells to 20 uM BDE-47 created inflammatory environment by stimulating secretion of IL-6 and IL-8. Antioxidant treatment significantly reduced the toxin stimulated IL-6 production, suggesting that secretion of pro-inflammatory molecules in this case may be facilitated through ROS. Enhanced production of ROS and superoxide anions is linked to increased risks of pre-term birth.

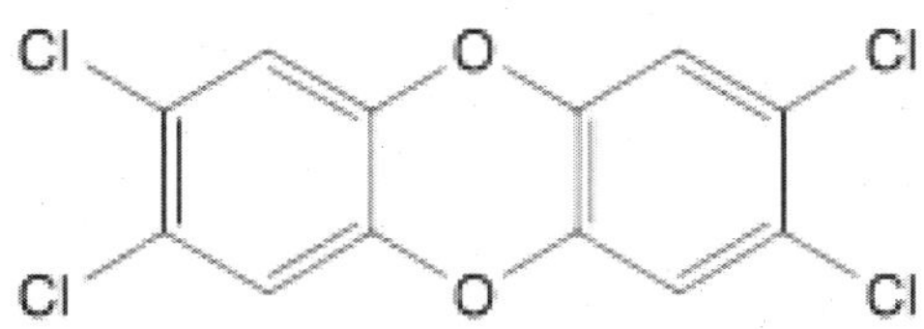

Figure 6. TCDD structure.

B. 2, 3, 7, 8-Tetrachlorodibenzo-p-Dioxin (TCDD)

TCDD is a colorless compound that consists of two benzene rings that are linked by two oxygen bridges (Figure 6). Chlorine atoms are attached at carbon atoms at positions 2, 3, 7, 8. The positions of chlorine atoms make TCDD one of the most toxic congeners of all polychlorinated dibenzodioxins.

TCDD has never been made commercially and it is produced unintentionally. According to Agency for Toxic Substances and Disease Registry (ATSDR), TCDD is a by-product of the manufacturing of polychlorinated phenol herbicides, paper and pulp bleaching, incineration of wastes, metal production, and fossil fuel, and wood combustion (ATSDR, 1998). TCDD is a persistent organic contaminant that is very lipophilic and accumulates in organisms. The most common route of exposure to this artificially made toxin is through ingestion of contaminated food. Ocean and fresh water fish has the highest amount of dioxin. Also, it has been found in meats, eggs, and milk produce such as butter, cheese, and ice cream (Schecter et al., 2011).

TCDD has been shown to cause many side effects on reproductive, developmental, and immune systems in animal models. The minimal risk level (MRL) of chronic TCDD exposure in humans was calculated to be 1 pg/kg/day and MRL of acute exposure was 0.2 ng/kg/day (ATSDR 2014). TCDD levels in breast milk samples evaluated in women in Japan were ranging from 0.1 pg/g lipid to 4.34 pg/g lipid (Nishijo et al., 2008).

TCDD acts as an EDC. Mechanism of its action is by activating aryl hydrocarbon receptor (AHR). Activation of AHR leads to modifications in gene transcription that may affect various tissues. The pollutant may increase risks of pregnancy complications by modulating immune system or hormonal levels.

Female/Male Fertility

TCDD may interfere with male and female fertility. Bruner-Tran et al., (2014) found that 50% of male mice (F1) exposed to 10ug/kg dose of the chemical in utero on day 15.5 of gestation were unable to impregnate their mating partners. Similarly, Ding et al., (2011) using the same experimental design observed that mating of control female mice with F1 males resulted in pregnancy in only 47% of animals. This effect was multigenerational, affecting F1-F3 male mice. Female mice exposed to the chemical in utero (F1) also exhibited reduced fertility in comparison to control animals. Mating F1 females with unexposed males resulted in 39% pregnancy (Ding et al., 2011). However, in the same study if both F1 partners were mated no pregnancy was observed. These two studies clearly indicate that high dose single contact to TCDD during pregnancy may affect reproductive health of offspring.

Pregnancy Outcomes (Miscarriage, Preeclampsia)

Pregnant rats exposed to 1.6 μg/kg body weight dose of TCDD on gestation day 15 exhibited delayed placental development by suppressing vascular remodeling (Wu et al., 2014). Modified placental vasculature may contribute to development of preeclampsia and fetal growth retardation. In the same study, animals exposed to high concentration of the toxin (8.0 μg/kg bw) had fetal demise. The toxin dramatically constricted labyrinth zone in placenta (in rodents area where nutrients, gases and wastes exchange) and reduced maternal blood flow. These modifications could contribute to the fetal death.

Many earlier studies did not observe negative effects of TCDD on fertility and only recently it was discovered that the negative effects caused by the toxin could be observed in the future generations (F1-F3). Similarly, prevalence of negative pregnancy outcomes in response to TCDD may be shown only in the future generation. More research needs to be done in this area. However, human studies will be difficult to perform because they will include evaluation of multiple generations. This will be very expensive and will take many years to get the answer.

Birth Outcomes (Pre-Term Birth, Low Birth Weight, SGA)

Epidemiologic studies demonstrate inconsistent results. A study performed in Japan found a negative correlation ($r = 0.369$, $p < 0.05$, $n = 41$) between newborn head circumference and TCDD levels in breast milk (Nishijo et al., 2008). Four mothers in the study delivered low birth weight infants. However, TCDD levels in breast milk samples obtained from those mothers were similar to the average concentration. The authors concluded that exposure to the toxin was not related to birth weight. In contrast, a retrospective Swedish study found that mothers who consumed high amount of fatty fish contaminated with polychlorinated dibenzo-p-dioxins and dibenzofurans had significantly higher chances of having low birth weight child or small for gestation age child (Rylander et al., 1995).

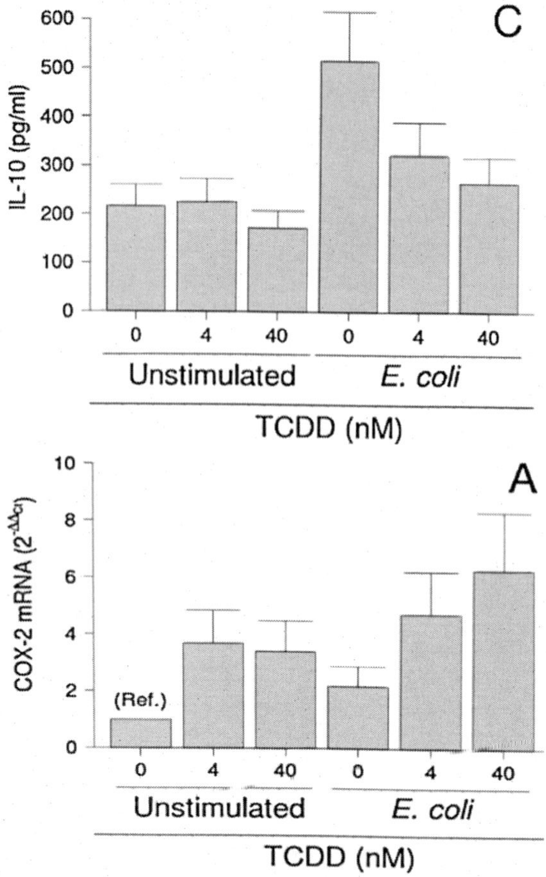

Figure 7. Effect of TCDD on placental IL-10 secretion (C) and COX-2 mRNA levels (A) in the presence or absence of infection. Adapted from Peltier et al., (2013).

Pregnant rats treated with low dose of the chemical (1.6 ng/kg and 8 ng/kg) on gestation day 15 had no observable effect on placenta or fetus weight (Wu et al., 2014). This suggests that high dose of the toxin is necessary to cause visible damage to the fetus. However, studies have shown that exposure to TCDD in utero may cause increased rates of pre-term birth and other complications later in life when an animal reaches reproductive age.

Ding et al., (2011) reported that if either parent exposed to TCDD in utero was mated with a control partner, rates of pre-term birth were significantly higher, 36% if exposed partner was female and 39% if exposed partner was male. Also, in the same study it was found that progesterone receptors in placenta and average placental and pup weight were decreased if either parent was exposed to TCDD. The authors suggested that adverse outcomes observed are due to inflammation and modulation of progesterone levels caused by the pollutant. Other scientists also observed that TCDD could increase inflammation. Peltier et al., (2013) reported that similar to PBDEs, TCDD might increase the risk of pre-term birth by enhancing placental inflammation in the presence of bacterial infection. Pre-exposing placental explants with either 4 nM or 40 nM doses of the toxin for 3 days significantly reduced *E.coli-* stimulated production of anti-inflammatory cytokine IL-10 (Figure 7C) and enhanced production of pro-inflammatory cytokine TNF-α.

In addition, the EDC may have more direct effect triggering pre-term birth by stimulating prostaglandins production, which is responsible for cervical softening and initiation of uterine contractions. As shown in Figure 7A, treatment of placental explants with 4 or 40 nM TCDD enhanced gene expression levels of cyclooxygenase 2 (COX-2), an enzyme responsible for production of prostaglandins. In the presence of *E. coli*, COX-2 mRNA levels were further upregulated in a dose dependent manner (Peltier et al., 2013).

C. Bis-Phenol-A (BPA)

BPA is a colorless solid that is poorly soluble in water. It is composed of two hydroxyphenyl groups (Figure 8).

BPA has been commercially used to manufacture plastics and resins since 1950s. For instance, polycarbonate is made by the reaction of BPA and phosgene. According to Environmental Protection Agency (EPA), each year more than 1 million pounds of the chemical are released to the environment. Polycarbonate is durable, easy to make, and highly transparent to visible light material. Polycarbonate is used to produce food containers, compact discs, impact-resistant safety equipment, and medical devices. Epoxy resins are used as protective coatings in food and beverage cans, bottle tops, and water pipes. BPA leaches from epoxy resins and contaminates food.

According to Food and Drug Administration (FDA) and EPA, daily exposure dose of 50 ug/kg/day is considered to be safe for humans. Measurable levels of BPA have been detected in neonatal and adult blood serum, amniotic fluid, placenta, breast milk, and urine. Average level of BPA circulating in the blood of adults is in the 1- to 3-ng/mL range (as sited in Vandenberg et al., 2009). BPA is found in concentrations ranging from 1 to 10 ng/mL in amniotic fluid and ranging from 1 to 104.9 ng/g in placenta tissues (Schonfelder et al., 2002). Possibly, humans are exposed to higher levels of the pollutant than estimated by FDA and EPA. Studies performed by Hunt et al., (2012) using rhesus monkeys as a model, demonstrated that 1-2 hrs after administration of 400 ug/kg/day dose of BPA orally, peak blood serum BPA levels were 2-5 ng/ml.

BPA accumulates in fat tissues. Exposure to the chemical has the most profound effect on health during fetal and neonatal development because fetal liver has reduced capacity to metabolize it (Elsby et al., 2001). BPA mechanism of toxicity is through binding to estrogen receptors (ER) and acting as an endocrine disruptor. BPA binds both ER α and ER β. Another pathwayα could be through altering androgen levels and thyroid hormone metabolism. In addition, BPA may initiate inflammatory response in adipose and liver tissues. It has been shown that BPA may disrupt the release of anti-inflammatory molecule adiponectin from fat tissues and high dose of BPA (50 mg/kg) can generate oxygen reactive species (ROS) in a rat liver and decrease expression of antioxidant genes (Hassan et al., 2012; Hugo et al., 2008). In addition, PBMC cells obtained from peripheral blood of healthy donors and cultured on a dish for 18 hours demonstrated that 10uM BPA can up-regulate 4 genes related to inflammation (Wens et al., 2013).

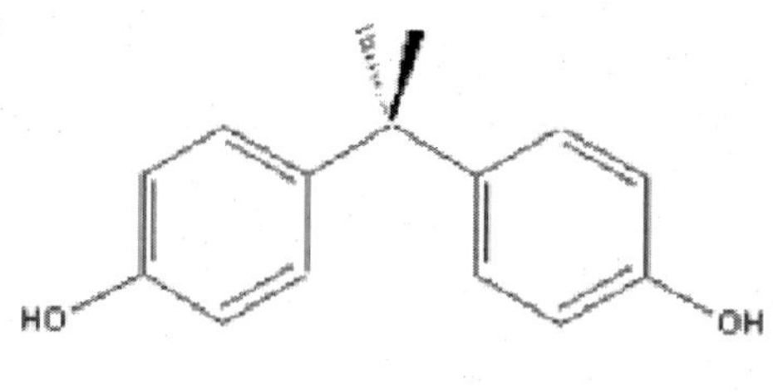

Figure 8. BPA structure.

FEMALE/MALE FERTILITY

Animal models have demonstrated that neonatal exposure of male rats or mice to the pollutant affects fertility. Male mice exposed to BPA had decreased sperm count and altered protein levels in the testis (Salian et al., 2009). It was suggested that the negative effects could be transferred to future generations by epigenetic changes. Rats exposed to 2.4 ug of BPA each day for the first five days of life had hypermethylated estrogen receptors α and β promoter region in testis (Doshi et al., 2011). Methylation patterns can regulate which genes are turned on or off. However, Goldstone et al., (2014) did not find any association between semen quality and urinal concentration of BPA in healthy humans.

Velez et al., (2015), evaluating 1742 pregnant women in Canada, concluded that high levels of BPA in the urine samples were not correlated with decreased fecundity, which was measured by the time it took these women to conceive (women with fertility problems, recruited for egg donation, birth control failure and male infertility were excluded from the study). Another study found that urinal BPA has been linked with fertility problems in women. Participants who had high concentrations of BPA had decreased number of mature oocytes and lower levels of estradiol in blood serum (Ehrlich et al., 2012). These findings are consistent with animal models. Administration of environmentally relevant doses of BPA to pregnant rhesus monkeys disrupted prophase I of meiosis I events and follicle formation in the fetal ovaries (Hunt et al., 2012). Also, female rats exposed to the chemical for seven days starting at postnatal day 1, exhibited reduction of gamete reserves (Rodriguez et al., 2010). Estrogen receptor beta expression and proliferation rates were significantly higher in recruited follicles obtained from animals that were treated with 20 mg/kg/day dose of BPA. The authors suggested that the early recruitment of primordial follicles could lead to earlier onset of menopause. However, rats exposed to 50 ug/kg/day dose had proliferation rates in recruited follicles similar to control animals.

ADVERSE PREGNANCY OUTCOMES (PREECLAMPSIA, MISCARRIAGE)

Leclerc et al., (2014) evaluating 58 pregnant women correlated high levels of BPA in placental tissue and preeclampsia. Another study has linked higher concentrations of the chemical in blood serum to increased chances of spontaneous miscarriage during first trimester of pregnancy (Lathi et al., 2014). The authors analyzed the pregnancy losses for chromosomal abnormalities and identified that women with BPA levels in the top 25% had

1.97 times increased risk for aneuploid miscarriage. Additional studies are necessary to confirm this finding. Similarly, Japanese scientists found a correlation between high serum BPA concentrations and recurrent miscarriage (Sigiura-Ogasawara et al., 2005). Women who experienced miscarriage had mean values of the pollutant equal to 2.59 ± 5.23 ng/ml and women who had normal pregnancy had BPA levels equal 0.77 ± 0.38 ng/ml.

Pregnant mice subcutaneously injected with BPA had significantly lower number of blastocyst implantation sites suggesting that the chemical may disrupt pregnancy in animal model (Berger et al., 2010). The authors also found that expression of estrogen and progesterone receptors was significantly reduced in these mice. The balance of progesterone and estrogen is important for the implantation and the researchers suggested that BPA can disrupt the balance.

Benachour and Aris (2009) suggested that exposure even to low concentrations of the toxin may modify placental cells, trophoblasts, which may explain why the chemicals can cause restricted growth of a fetus. The authors found out that exposing primary cytotrophoblast cells isolated from fresh human placenta to low levels of BPA ranging from 0.001 to 1000 ug/ml increased rates of programmed cell death (apoptosis) in a dose dependent manner. In addition, exposure to very low doses of BPA for 24 hours increased rates of necrosis and exposure to higher doses decreased necrosis (Benachour & Aris, 2009). Chronic exposure to the chemical may cause more dramatic changes to the cells affecting placental development, which in turn could lead to pregnancy complications such as preeclampsia, intrauterine growth restriction, premature birth, and miscarriage.

Adverse Birth Outcomes

A study assessing levels of BPA in placentas in the Southeastern U.S has detected BPA in every placenta analyzed with concentrations ranging from 0.64 to 39.7 ng/g/wet weight (Troisi et al., 2014). The same study concluded that low birth weight infants (<2500 g) and small for gestational age infants had significantly increased levels of BPA in placentas. Similarly, Chou et al., (2011) correlated high BPA levels in maternal plasma with increased odds of giving birth to a low birth weight neonate or SGA infant. Parental exposure to the chemical at work was linked with neonatal reduced birth weight. Birth weight of infants born from exposed mothers was on average 176 g less in comparison with neonates from mothers without BPA exposure in the workplace (Miao et al., 2011).

In contrast, Wolff et al., (2008) found no correlation between BPA levels in urine samples obtained during 3rd trimester and birth weight. Another study performed in Korea found that the average concentration of BPA in urine samples of women during third trimester pregnancy was 1.08 ug/L (ranging from not detectable to180.6 ug/l) (Lee et al., 2014). The authors found a positive correlation between BPA concentration and weight of male newborns only, which is not in agreement with previously described study. The explanation could be that urine test was done only one time and assessed levels of BPA represented only the levels present at the day of test not throughout the pregnancy. Perhaps, measuring BPA levels in urine is not a good indicator of exposure levels. Nevertheless, there are many contradictory studies regarding how BPA affects birth weight. This could reflect differences in population, methodology and design of the study, and timing of exposure to BPA used among researchers.

One small nested control study in Mexico evaluated risk of pre-term birth in relation to urinary levels of BPA. The researchers found that women who had higher levels of BPA were more likely to deliver prematurely (Cantonwine et al., 2010). The average concentration of BPA in the urine for women who delivered premature was 1.84 ug/l and for women who delivered at term was 0.97 ug/l. Further studies are needed to clarify the effect of BPA exposure on risk of preterm, SGA, and low birth weight.

D. Bis 2-Ethylhexyl Phthalate (DEHP)

Phthalates are used in plastics industry as additives to soften polyvinyl chloride products. Examples of products that contain phthalates are insulators of electrical cables, clothes (leather like materials), car interiors, toys, personal care products, medical devises, and pharmaceuticals (Schettler, 2006). Bis 2-ethylhexyl phthalate (DEHP) (Figure 9) is the most common phthalate used in plastics industry, accounting for more than 50%. It is colorless viscous liquid, not soluble in water. DEHP has low volatility levels and low cost. Phthalates can leach from plastics and contaminate food and water.

Phthalates have been linked to alterations of thyroid and reproductive hormone levels, and lipid metabolism (Johns et al., 2015). Sathyanarayana et al., (2014) reported that high levels of urinary DEHP metabolites were associated with significantly lower serum testosterone levels among pregnant women. According to FDA, tolerable intake values for DEHP are 40 ug/kg/day for oral exposure. A study performed in Germany estimated that humans are exposed to higher concentrations of DEHP than it is commonly thought (Koch et al., 2003). The authors calculated DEHP levels from 85 urine samples and concluded that 12% of the participants had levels higher than the tolerable value. The average concentration of DEPH in their study was 13.8 ug/kg/day body weight. Other scientists reported the average DEHP concentration in urine samples to be between 2.32 to 12 ug/kg/day, other phthalates appear at lower concentrations (as cited in Kay et al., 2013). However, exposure to multiple phthalates in addition to DEHP may have additive effect in the body.

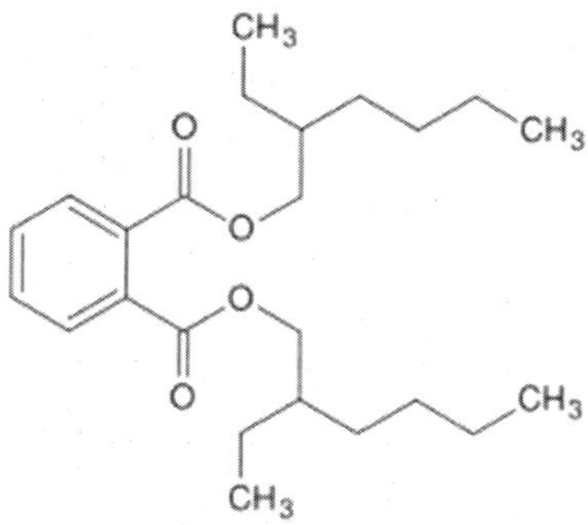

Figure 9. DEHP structure.

FEMALE/MALE FERTILITY

Canadian study measuring the time it took 1597 women to get pregnant concluded that presence of phthalates were not associated with increased time to conception (Velez et al., 2015). The authors measured only 11 phthalate metabolites in the urine samples obtained one time during first trimester. The concentrations of chemicals may not be the same during the preconception stages.

Animal models demonstrated that exposure to phthalates may have negative effect on oocyte development. Rats exposed in utero to diisobutyl phthalate exhibited oocyte degeneration and abnormal follicles (Ray et al., 2012). One study demonstrated that exposure to DEHP, similarly to TCDD, may cause negative effects in later generations. Doyle et al., (2013) demonstrated that exposing mice in utero for 7 days (embryonic days 7-14) to

500 mg/kg per day DEHP (F1) produced F2, F3, and F4 generation of male mice that had significant decrease in sperm count and motility. The negative results were similar if exposed animals were in maternal, in paternal, or both lineages.

PREGNANCY OUTCOMES

Animal studies suggest that high doses of phthalates may cause pregnancy complications. All female mice that were fed with 500 mg/kg of DEHP for 8 weeks had a miscarriage; however, a log lower dose did not cause severe adverse effects on pregnancy in these animals (Schmidt et al., 2012). In the same study, mice treated with lower doses of DEHP 5 mg/kg and 0.05 mg/kg in utero, had higher blastocyst degeneration rates in comparison to control mice, 29%, 28%, and 8%, respectively. However, the results were not statistically significant.

In humans, high levels of DEPH in urine were associated with shorter gestation by 4.2 days and increased chances of early pregnancy loss (Toft et al., 2012; Weinberger et al., 2014).

BIRTH OUTCOMES

Women who had high levels of DEHP and its metabolites in urine samples (n = 1693) had significantly higher risks of preterm and spontaneous preterm birth (Ferguson et al., 2014). 95% of these samples were positive for phthalates. In the same pool of samples, oxidative stress markers 8-isoprostane and 8-OHdG were evaluated. Presence of phthalates was associated with increased levels of these markers (Ferguson et al., 2014). This finding implies that besides acting as endocrine disruptors, the chemicals may cause damage by initiating oxidative stress mechanisms. The same of group of scientists correlated elevated levels of the chemicals in urine samples obtained during first trimester of pregnancy and slightly higher chances of preeclampsia and intrauterine growth restriction (Ferguson et al., 2014). Similarly, Zhao et al., (2014) observed that presence of elevated levels of phthalates in urine collected during third trimester was associated with intrauterine growth restriction. Wolff et al., (2008) found no association between urinary levels of

DEPH and its metabolites measured during 3rd trimester of pregnancy and birth weight.

E. Phytoestrogens

Phytoestrogens are naturally occurring non-steroidal compounds that have estrogenic properties. Examples of products that contain phytoestrogens are soymilk, tofu, and other soy containing products, also flax, sunflower, and sesame seeds. Several studies have shown that phytoestrogens may have endocrine disrupting effects due to their similarity to estrogen (Figure 10). The compounds can bind to estrogen receptors alfa and beta and activate them. The major types of phytoestrogens are isoflavones, lignans and coumestans. Isoflavones daidzein and genistein are the most studied.

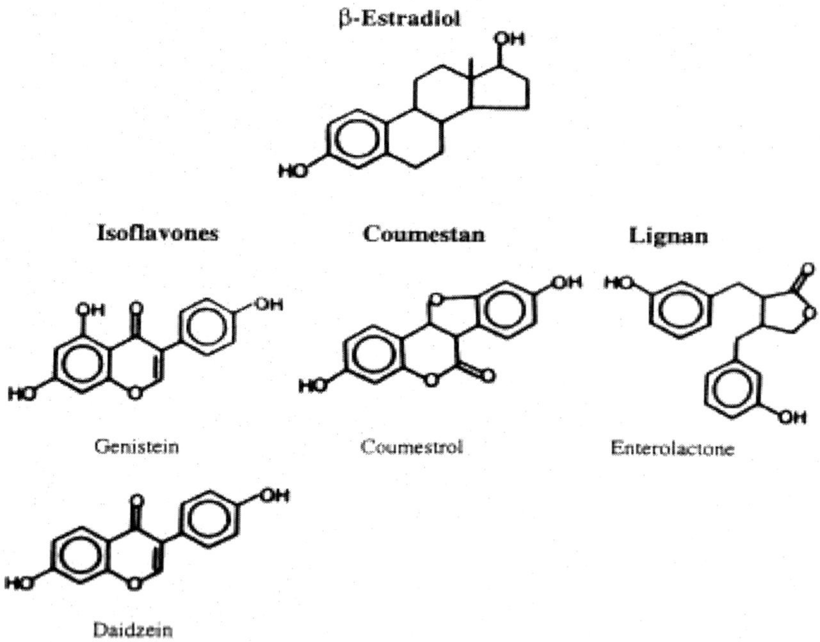

Figure 10. Structures of estrogen and common phytoestrogens.

Isoflavones may have a positive influence on cardiovascular system, cancers, and diabetes; however the positive results observed are marginal and only few studies were done. To draw a conclusion about their effect on health further investigation is necessary. One of the proposed mechanisms of their action is by increasing production of antioxidant defense enzymes and nitric-oxide synthase. In fish, genistein, flavones, and other phytoestrogens inhibit ovarian aromatase activity, an enzyme that converts testosterone to estrogen (Pelissero et al., 1996). However, in the same study daidzein and some other phytoestrogens did not affect aromatase activity. Daidzein and genistein are detectable in serum, amniotic fluid, breast milk, and cord blood (Jarrell et al., 2012). Degen et al., (2002) demonstrated that daidzein could transfer across the placenta in a rat model; therefore, it may potentially affect pregnancy and outcomes.

FEMALE/MALE FERTILITY

Genistein may negatively affect female fertility. Female mice injected with genistein as neonates on days 1-5 after birth, exhibited decreased fertility in a dose dependent manner (Jefferson et al., 2005). Their ability to get pregnant was evaluated at 2, 4, and 6 months of age. The decrease in fertility was especially noticeable at 6 month of age. Mice treated with the highest dose of the compound 100 ug per animal, could not carry out the pregnancy. They had fewer implantation sites and higher reabsorption rates. Also, it was shown that genistein may have an adverse impact on oocytes. Oocytes treated with genistein *in vitro*, had reduced ability to be fertilized (Chan, 2009). Young female rats that were exposed to soy rich diet throughout their lifetime had increased number of cystic follicles, elevated blood glucose levels, and irregular estrus cycle (Patisaul et al., 2014). All of these symptoms are characteristics of polycystic ovary syndrome, which may cause fertility problems.

The effect of phytoestrogens on male fertility is unclear. Fielden et al., (2003) found no negative impact of genistein on sperm quality of male mice exposed to the compound in utero and during lactation. Male mice exposed to the highest dose used in the study 10mg/kg/day, demonstrated increased fertilizing ability of their epididymal sperm by 17%. The other study, showed the opposite results. Male mice that were fed diet rich in phytoestrogens throughout their lives had decreased epididymal sperm count by 25% and reduced fertility (Cederroth et al., 2010).

Pregnancy Outcomes (Miscarriage, Preeclampsia)

Isoflavones may alter ratio of pregnancy hormones in human placenta. Treating trophoblast cells isolated from term placentas for 24 hours with genistein and daidzein increased production of estrogen and reduces secretion of progesterone and hCG (Richter et al., 2009). It implies that presence of high doses of these compounds may have a negative impact on pregnancy. Very few studies are available to make a definite conclusion. More studies are needed to evaluate risks of developing preeclampsia and miscarriage in response to consumption of phytoestrogens.

Birth Outcomes (Pre-Term Birth, Low Birth Weight, SGA)

One study found no correlation between phytoestrogens present in amniotic fluid, blood serum, or cord serum and negative pregnancy outcomes. About 93% of all pregnant women who participated in the study had detectable levels of either genistein or daidzein in amniotic fluid with mean concentration equal to 0.7 ng/ml (Jarrell et al., 2012). The average concentration of daidzein in blood serum was 2.88 ng/ml and the levels of genistein were 5.18 ng/ml. Presence of these phytoestrogens did not affect pre-term birth rates, incidence of preeclampsia, or low birth weight. The study of a bigger in size is needed to make definite conclusions.

Conclusion

Exposure to PBDEs, TCDD, BPA, phthalates and phytoestrogens may negatively impact female and male fertility, pregnancy outcomes, and birth outcomes. It appears that phytoestrogens have less harmful effect on reproductive health than man-made chemicals. However, there are many gaps in the data and some results are not clear.

In vitro experiments show that some toxins have the ability to modify cells by altering apoptosis and necrosis rates and others may cause oxidative stress and enhance inflammation reaction. Animal studies demonstrate that exposure to the toxins in utero may adversely affect reproductive health for the lifetime of the organism, increasing chances of reduced fertility, negative pregnancy outcomes, and poor birth outcomes. These changes are irreversible and maybe be transgenerational. Animal models provide information about acute impacts of the pollutants on reproductive health. There are few studies that use environmentally relevant doses of the pollutants to assess their effects on health. Different doses of the chemicals may have various impacts on health. Another challenge using animal models is that results received from long - term studies may not be applied directly to humans because of differences in length of gestation, hormone levels during pregnancy, placenta protein products, and metabolism of the chemicals.

Epidemiological studies often correlate exposure to high concentrations of the chemicals with increased risks of adverse outcomes. However, there are many challenges associated with studying effects of environmental pollutants on human fertility, pregnancy, and outcomes. First of all, other factors such as physical activity levels, smoking, education, existing medical conditions, diet, and social-economic status should be taken into account because these factors may influence the successes of reproduction. In addition, we are exposed to many chemicals at the same time. It is unclear how mixture of the toxins will affect our health. Will the effects be similar to each individual toxin or altered in some way? Wang et al., (2014) studied effect of combined genistein and BPA treatment on mammary gland development. Young rats, 2 to 20 days old, were exposed to the chemicals via milk from their mothers. The combined treatment significantly induced cell proliferation and reduced programmed cell death, apoptosis at 21 days, and at 50 days of age the effect was the opposite with decreased cell proliferation rates and increased apoptosis. The results were similar to the genistein exposure alone. BPA and genistein both are estrogenic compounds and it could be expected that their effects may be in an additive way, but this experiment demonstrated that different pathways could be activated. However, another study, evaluating co-treatment of PCBs and PBDEs on rats during gestation and lactation found that the chemicals had additive effects on thyroid hormone levels (Miller et al., 2012). Recent studies suggest that the environmental toxins TCDD and BPA may have adverse effect on several generations by changing DNA methylation patterns. Possibly, other toxins such as DEHP, PBDE, and heavy metals could act in a similar way. Furthermore, effects of chronic exposure to low levels of chemicals on health may be concealed. All of these issues add to the complexity of the problem. Nevertheless, despite uncertainties and knowledge gaps, the available evidence suggests that limiting exposure to these chemicals especially during pregnancy is advisable.

REFERENCES

Agarwal, A., Aponte-Mellado, A., Premkumar, B. J., Shaman, A. & Gupta, S. (2012). The effects of oxidative stress on female reproduction: a review. *Reproductive Biology and Endocrinology*, *10*, 49.

Agency for Toxic Substances and Disease Registry (ATSDR) (2007). Toxicological Profile for Lead. Retrieved from http://www.atsdr.cdc.gov/toxprofiles/tp.asp?id=96&tid=22.

Alaee, M., Arias, P., Skodin, A. & Bergman, A. (2003). An overview of commercially used brominated flame retardants, their application, their use patterns in different countries/regions and possible modes or release. *Environment International*, *29*, 683-689.

Arredondo, F. & Noble, L. S. (2006). Endocrinology of recurrent Pregnancy Loss. *Seminars in Reproductive Medicine*, *24*(1), 033-039.

Avanth, C. V. & Vintzileos, A. (2006). Maternal-fetal conditions necessitating a medical intervention resulting in preterm birth. *American Journal of Obstetrics and Gynecology*, *195*, 1557-63.

Augustowska, K., Gregoraszczuk, E. L., Milewicz, T., Krzysiek, J., Grochowalski, A. & Chrzaszcz, R. (2003). Effects of dioxin (2, 3, 7, 8-TCDD) and PCDDs/PCDFs congeners mixture on steroidogenesis in human placenta tissue culture. *Endocrine Regulations*, *37*, 11e9.

BSEF. (2000). An Introduction to Bromine Brussels: Bromine Science and Environmental Forum. Retrieved from http://www.bsefsite.com/docs/bromine.pdf.

Benachour, N. & Aris, A. (2009). Toxic effects of low doses of Bisphenol-A on human placental cells. *Toxicology and Applied Pharmacology.*, *241*(3), 322-328.

Berger, R. G., Foster, W. G. & deCatanzaro, D. (2010). Bisphenol A exposure during the period of blastocyst implantation alters uterine morphology and perturbs measures of estrogen and progesterone receptor expression in mice. *Reproductive Toxicology*, *30*(3), 393, 400.

Birth weight & Gestation. (2014). Centers for Disease Control and Prevention. Retrieved from http://www.cdc.gov/nchs/data/nvsr/nvsr64/nvsr64_06.pdf.

Bruner-Tran, K. L., Ding, T., Yeoman, K. B., Archibong, A., Arosh, J. A. & Osteen, K. G. (2014). Developmental exposure of mice to dioxin promotes transgenerational testicular inflammation and an increased risk of preterm birth in unexposed mating partners. *PLoS One*, *9*(8), e105084.

Burton, G. L., Yung, H. W., Cindrova-Davies, T. & Charnock-Jones, D. S. (2009). Placental endoplasmic reticulum stress and oxidative stress in the

, P. R. S., Coburn, G., Moser, V. C., MacPhail, R. C., Fenton, S. E., , T. E., Rayner, J. L., Kannan, K. & Birnbaum, L. S. (2010). opmental exposure to a commercial PBDE mixture, DE-71: behavioral, hormonal, and reproductive effects. *Toxicological ces*, *116*(1), 297-312.

, S. N., Talsness, C. E., Grote, K. & Chahoud, I. (2005). lopmental exposure to low-dose PBDE-99: effects on male fertility neurobehavior in rat offspring. *Environmental Health Perspectives*, 2), 149–154.

B., Liebert, C. A., Brookfield, K. F., Taylor, J. A., vom Saal, F. S., moto, V. Y. & Baker, V. L. (2014). Conjugated bisphenol A in rnal serum in relation to miscarriage risk. *Fertility and Sterility*, (1), 123-128.

F., Dubois, M. F. & Aris, A. (2014). Maternal, placental and fetal osure to bisphenol A in women with and without preeclampsia. *pertensive Pregnancy*, *33*(3), 341-348.

E., Park, H., Hong, Y. C., Ha, M., Kim, Y., Chang, N., Kim, B. N., n, Y. J., Yu, S. D. & Ha, E. H. (2014). Prenatal bisphenol A and birth comes: MOCEH (mothers and children's Environmental Health) study. *giene and Environmental Health*, *217*, 328-334.

, R. (2012). Oxidative stress damage as a detrimental factor in preterm rth Pathology. *Frontier Immunology*, *5*, 1-14.

M., Yuan, W., Zhu, G., He, X. & Li, D. K. (2011). In utero exposure to sphenol A andits effect on birth weight of offspring. *Reproductive oxicology*, *32* (1), 64-68.

, M. F., Chernyak, S. M., Domino, S. E., Batterman, S. A. & Loch- aruso, R. (2012). Concentration and speciation of polybrominated iphenyl ethers in human amniotic fluid. *Science of the Total Environment*, 417-418, 294-298.

r, V. M., Sanchez-Morrissey, S., Brosch, K. O. & Seegal, R. F. (2012). Developmental coexposure to polychlorinated bisphenyls and polybrominated diphenyl ethers had additive effects on circulation thyroxine levels in rats. *Toxicological Sciences*, *127*(1), 76-83.

imal Risk Levels (MRLs). Agency for Toxic Substances and Disease Registry 2015. Retrieved from www.atsdr.cdc.gov/mrls/index.html# bookmark02.

ren, M., Mose, T., Mathiesen, L. & Knudsen, L. E. (2007). The human placenta—an alternative for studying foetal exposure. *Toxicology In Vitro*, *21*, 1332-1340.

pathophysiology of unexplained intrauterine growth restriction and early onset of preeclampsia. *Placenta*, *30* (S), 43-48.

Cantonwine, D., Meeker, J. D., Hu, H., Sanchez, B. N., Lamadrid-Figueroa, H., Mercado-Garcia, A., For then berry, G. Z., Calafat, A. M. & Tellez-Rojo, M. M. (2010). Bisphenol A exposure in Mexico City and risk of prematurity: a pilot nested case control study. *Environmental Health*, *9*, 62.

Caserta, D., Maranghi, L., Mantovani, A., Marci, R., Maranghi, F. & Moscarini, M. (2008). Impactof Endocrine disruptor chemicals in Gynaecology. *Human Reproduction Update*, *14*(1), 59-72.

Cederroth, C. R., Zimmermann, C., Beny, J. L., Schaad, O., Combepine, C., Descombes, P., Doerge, D. R., Pralong, F. P., Vassalli, J. D. & Nef, S. (2010). Potential detrimental effects of a phytoestrogen-rich diet on male fertility in mice. *Molecular Cell Endocrinology*, *321*(2), 152-160.

Centers for Disease Control and Prevention (CDC), (2014). Reproductive Health. Retrieved from http://www.cdc.gov/reproductivehealth/infertility/.

Chan, W. H. (2009). Impact of genistein on maturation of mouse oocytes, fertilization, and fetal development. *Reproductive Toxicology*, *28*, 52-58.

Chou, W. C., Chen, J. L., Lin, C. F., Chen, Y. C., Shih, F. C. & Chuang, V. Y. (2011). Biomonitoring of bisphenol A concentrations in maternal and umbilical cord blood in regard to birth outcomes and adipokine expression: a birth cohort study in Taiwan. *Environmental Health*, *10*(94), 1-10.

Degen, G. H., Janning, P., Diel, P., Michna, H. & Bolt, H. M. (2002). Transplacental transfer of the phytoestrogen daizein in DA/Han rats. *Archives of Toxicology*, *76*(1), 23-29.

Doshi, T., Mehta, S. S., Dighe, V., Balasinor, N. & Vanage, G. (2011). Hypermethylation of estrogen receptor promoter region in adult testis of rats exposed neonatally to bisphenol A. *Toxicology*, *289*(2-3), 74-82.

Doyle, T. J., Bowman, J. L., Windell, V. L., McLean, D. J. & Kim, K. H. (2013). Transgenerational effects of di-(2-ethylhexyl) phthalate on testicular germ cell associations and spermatogenial stem cells in mice. *Biology of Reproduction*, *88*(5), 112.

Du, P., Li, Z., Du, L., Zhang, H., Zhou, Y., Sun, W., Xiao, X., He, Y., Sun, B., Yu, Y. & Chen, D. (2015). The effects of PBDE-209 on placental ET-1 and eNOS expression and the birth of offspring. *International Journal of Developmental Neuroscience*, *43*, 86-91.

Ehrlich, S., William, P. L., Missmer, S. A., Flaws, J. A., Ye, X., Calafat, A. M., Petrozza, J. C., Wright, D. & Hauser, R. (2012). Urinary bisphenol A concentration and early reproductive health outcomes among women undergoing IVF. *Human Reproduction*, *27*(12), 3583-3592.

Elsby, R., Maggs, J. L., Ashby, J. & Park, B. K. (2001). Comparison of the modulatory effects of human and rat liver microsomal metabolism on the estrogenicity of bisphenol A: implication for extrapolation to humans. *Pharmacology and Experimental Therapeutics*., *297* (1), 103-113.

Environmental Protection Agency, (2014). Assessing and Managing Chemicals under TSCA. Polybrominated diphenyl ethers (PBDEs) action plan summary. Retrieved from http://www.epa.gov/assessing-and-managing-chemicals-under-tsca/polybrominated-diphenyl-ethers-pbdes.

Ferguson, K. K., McElrath, T. F., Chen, Y. H., Mukherjee, B. & Meeker, J. D. (2015). Urinaryphthalate metabolites and biomarkers of oxidative stress in pregnant women: a repeated measures analysis. *Environmental Health Perspectives Mar*, *123*(3), 210-6.

Ferguson, K. K., McElrath, T. F., Ko, Y. A., Mukherjee, B. & Meeker, J. D. (2014). Variability of urinary phthalate metabolite levels across pregnancy and sensitive windows of exposure for the risk of pre-term birth. *Environmental International*, *70*, 118-124.

Ferguson, K. K., McElrath, T. F. & Meeker, J. (2014). Environmental phthalate exposure and preterm birth. *American Medical Association of Pediatrics*, *168*(1), 61-68.

Fielden, M. R., Samy, S. M., Chou, K. C. & Zacharewski, T. R. (2003). Effect of human dietary exposure levels of genistein during gestation and lactation on long term reproductive development and sperm quality in mice. *Food and Chemical Toxicology*, *41*(4), 447-454.

Goldenberg, R. L., Culhane, J. F., Iams, J. D. & Romero, R. (2008). Epidemiology and Causes of Preterm Birth. *Lancet*, *371*, 75-84.

Goldstone, A. E., Chen, Z., Perry, M. J., Kannan, K. & Louis, G. M. (2014). Urinary bisphenol A and semen quality, the LIFE study. *Reproductive Toxicology*, *51*C. 7-13.

Harley, K. G., Marks, A. R., Chevrier, J., Bradman, A., Sjodin, A. & Eskenazi, B. (2010). PBDE concentrations in women's serum and fecundability. *Environmental Health Perspectives*, *118*(5), 669-704.

Hassan, Z. K., Elobeid, M. A., Virk, P., Omer, S. A., ElAmin, M., Daghestani, M. H. & Alolayan, E. M. (2012). Bisphenol A induces hepatotoxicity through oxidative stress in rat model. *Oxidative Medicine and Cellular Longevity*, 2012, 194829.

Environmental Protection Agency (EPA), PCBsWashington (DC). Retrieved from hazard/tsd/pcbs/pubs/effects.htm.

Hugo, E. R., Brandebourg, T. D., Woo, J. G., Ben-Jonathan, N. (2008). Bisphenol A at e inhibits adiponectin release from human adipocytes. *Environmental Health Perspectiv*

Hunt, P. A., Lawson, C., Gieske, M., Murdoch Hassold, T. & VandeVoortb, C. A. (2012) oogenesis and follicle formation in the fetal o *Proceedings of the National Academy of Scienc*

Jarrell, J., Foster, W. G. & Kinniburgh, D. W. human pregnancy. *Obstetrics and Gynecol* 850313.

Jefferson, W. N., Padilla-Banks, E. & Newbold, R. R on female development and reproduction in CD-1 exposure tothe phytoestrogen genistein at environ *Biology of Reproduction*, *75*(4), 798-806.

Johns, L. E., Ferguson, K. F., Soldin, O. P., Cant Gonzalez, L. O., Anzalota Del Toro, L. V., C Alshawabkeh, A. N., Cordero, J. F. & Meeker, phthalate metabolites in relation to maternal se hormone levels during pregnancy: a longitudinal a *Biology & Endocrinology*, *13*, 4.

Johnson, P. I., Altshul, L., Cramer, D. W., Missmer, Meeker, J. D. (2012). Serum and follicular flui polybrominated diphenyl ethers and *in-vitro* fer *Environmental International*, *45*, 9–14.

Kalkunte, S., Nevers, T., Norris, W. E. & Sharma, S. (2011) protective role in preeclampsia. *Journal of Reproducti* (2), 165-169.

Kay, V. R., Chambers, C. & Foster, W. G. (2013). developmental effects of phthalate diesters in females. *Toxicology*, *43*(3), 200-219.

Koch, H. M., Drexler, H. & Angerer, J. (2003). An estimat intake of di (ethylhexyl) phthalate (DEHP) and other p general population. *International Journal of Hygiene and Health*, *206* (2), 77-83.

Kodavanti Stoke Deve neurc *Scier*

Kuriyam Deve and *113*

Lathi, R Fuji mat *102*

Leclerc exp *Hy*

Lee, B Ki ou *H*

Meno bi

Miao, b *T*

Mille ((

Mill

Mi

My

Nishijo, M., Tawara, K., Nakagawa, H., Honda, R., Kido, T., Nishijo, H. & Saito, S. (2008). 2, 3, 7, 8-Tetrachlorodibenzo-p-dioxin in maternal breast milk and newborn head circumference. *Exposure Science & Environmental Epidemiology*, *18*(3), 246-251.

Patisaul, H. B., Mabrey, N., Adewale, H. B. & Sullivan, A. W. (2014). Soy but not bisphenol A (BPA) induces hallmarks of polycystic ovary syndrome (PCOS) and related metabolic co-morbidities in rats. *Reproductive Toxicology*, *49*, 209-218.

Park, H. R., Kamau, P. W. & Loch-Caruso, R. (2014). Involvement of reactive oxygen species in brominated diphenyl ether-47-induced inflammatory cytokine release from human extravillous trophoblasts *in vitro*. *Toxicology and Applied Pharmacology*, *274*, 283-292.

Pelissero, C., Lenczowski, M. J., Chinzi, D., Davail-Cuisset, B., Sumpter, J. P. & Fostier, A. (1996). Effects of flavonoids on aromatase activity, an *in vitro* study. *Steroid Biochemistry & Molecular Biology*, *57*(3-4), 215-223.

Peltier, M. R., Arita, Y., Klimova, N. G., Gurzenda, E. M., Koo, H. C., Murthy, A., Lerner, V. & Hanna, N. (2013). 2, 3, 7, 8-Tetrachlorodibenzo-p-dioxin (TCDD) enhances placental inflammation. *Reproductive Immunology*, *98*, 10-20.

Peltier, M. R., Klimova, N. G., Arita, Y., Gurzenda, E. M., Murthy, A., Chawala, K., Lerner, V., Richardson, J. & Hanna, N. (2012). Polybrominated diphenyl ethers enhance the production of proinflammatory cytokines by the placenta. *Placenta*, *33*(9), 745-749.

Peltier, M. R., Koo, H. C., Getahun, D. & Menon, R. (2014). Does exposure to flame retardants increase the risk for preterm birth? *Reproductive Immunology*, *107*, 20-5.

Pomini, F., Caruso, A. & Challis, J. R. (1999). Interleukin-10 modifies the effects of interleukin-1beta and tumor necrosis factor-alpha on the activity and expression of prostaglandin H synthase-2 and the NAD+-independent 15-hydroxyprostaglandin dehydrogenase in cultured term human villous trophoblast and chorion trophoblast cells. *Clinical Endocrinology & Metabolism*, *84*, 4645-4650.

Poppe, K., Velkeniers, B. & Glinoer, D. (2008). The role of thyroid autoimmunity in fertility and pregnancy. *Nature Clinical Practice Endocrinology & Metabolism*, *4*, 394-405.

MedlinePlus, (2014). Preeclampsia. Bethesda, MD: National Library of Medicine (US). [accessed on November 2014]. Retrieved from http:// www.nlm.nih.gov/medlineplus/[accessed on November 2014]. Preterm birth in the US data. 2015. March of Dimes. Retrieved fromhttp://www.marchofdimes.org/mission/prematurity-reportcard.aspx.

Rajapakse, N., Silva, E. & Kortenkamp, A. (2002). Combining xenoestrogens at levels below individual no-observed effect concentration dramatically enhances steroid hormone action. *Environmental Health Perspectives*, *110*, 917-921.

Ray, B., D'Souza, A. S., Kumar, V., Pugazhandhi, B., D'Souza, M. R., Nayak, D., Sushma, R. K., Shetty, P., Singh, H., Krisna, L., et al., (2012. Ovarian development in Wistar rat treated prenatally with single dose diisobutyl phthalate. *Bratislavske Lekarske Listy*, *113*(10), 577-582.

Regev, R. H. & Reichman, B. (2004). Prematurity and intrauterine growth retardation--double jeopardy? *Clinical Perinatology*, *31*(3), 453-73.

Roberts, C. P. & Murphy, A. A. (2000). Endocrinopathies Associated with Recurrent Pregnancy Loss. *Seminars in Reproductive Medicine*, *18*(4), 357-362.

Rodríguez, H. A., Santambrosio, N., Santamaría, C. G., Muñoz-de-Toro, M. & Luque, E. H. (2010). Neonatal exposure to bisphenol A reduces the pool of primordial follicles in the rat ovary. *Reproductive Toxicology Dec*, *30*(4), 550-7.

Richter, D. U., Mylonas, I., Toth, B., Scholz, C., Briese, V., Friese, K. & Jeschke, U. (2009). Effects of phytoestrogens genistein and daidzein on progesterone and estrogen (estradiol) production of human term trophoblast cells *in vitro*. *Gynecological Endocrinology*, *25*(1), 32-38.

Rylander, L., Stromberg, U. & Hagmar, L. (1995). Decreased birth weight in infants born to women with a high dietary intake of fish contaminated with persistent organochlorine compounds. *Scandinavian Journal of Environmental Health*, *21*(5), 368-375.

Salian, S., Doshi, T. & Vanage, G. (2009). Neonatal exposure of male rats to bisphenol A impairs fertility and expression of sertoli cell junctional protein in the testis. *Toxicology*, *265*, 56-67.

Sathyyanarayana, S., Barrett, E., Butts, S., Wang, C. & Swan, S. H. (2014). Phthalate and Reproductive Hormone Concentration in Pregnancy. *Reproduction*, *147*(4), 401-409.

Schecter, A., Smith, S., Colacino, J., Malik, N., Opel, M., Paepke, O. & Birnbaum, L. (2011). Contamination of U.S. butter with polybrominated diphenyl ethers from wrapping paper. *Environmental Health Perspectives.*, *119*(2), 151-4.

Schettler, T. (2006). Human exposure to phthalates via consumer products. *International Journal of Andrology*, *29*(1), 134-139.

Schmidt, J. S., Schaedlich, K., Fiandanese, N., Pocar, P. & Fischer, B. (2012). Effects of Di(2-ethylhehyl) phthalate (DEHP) on female fertility and adipogenesis in C3H/N mice. *Environmental Health Perspectives*, *120*, 1123-1129.

Schonfelder, G., Wittfoht, W., Hopp, H., Talsness, C. E., Paul, M. & Chahoud, I. (2002). Parent bisphenol A accumulation in the human maternal-fetal-placental unit. *Environmental Health Perspectives*, *110*, A703-A707.

Sigiura-Ogasawara, M., Ozaki, Y., Sonta, S. I., Makino, T. & Suzumori, K. (2005). Exposure to bisphenol A is associated with recurrent miscarriage. *Human Reproduction*, *20*(8), 2325-23-29.

Talsness, C. E., Kuriyama, S. N., Sterner-Kock, A., Schnitker, P., Grande, S. W., Shakibaei, M., Andrade, A., Grode, K. & Chahoud, I. (2008). In utero and lactational exposures to low doses of ploybrominated diphenyl ether-47 alter the reproductive system and thyroid gland of female rat offspring. *Environmental Health Perspectives*, *116*(3), 308-314.

Tjoa, M. L., van Vugt, J. M. G., Go, A. T. J. J., Blankenstein, M. A., Oudejans, C. B. M. & van Wijk, I. J. (2003). Elevated C-reactive protein levels during first trimester of pregnancy are indicative of preeclampsia and intrauterine growth restriction. *Reproductive Immunology*, *59*(1), 29-37.

Toft, G., Jonsson, B. A. G., Lindh, C. H., Jensen, T. K., Hjollund, N. H., Vested, A. & Bonde, J. P. (2012). Association between pregnancy loss and urinary phthalate levels around the time of conception. *Environmental Health Perspectives*, *120*(3), 458-463.

Troisi, J., Mikelson, C., Richards, S., Symes, S., Adair, D., Zullo, F. & Guida, M. (2014). Placental concentrations of bisphenol A and weight from births in the Southeastern U. S. *Placenta.*, *35* (11), 947-952.

Vandenberg, L. N., Maffini, M. V., Sonnenschein, C., Rubin, B. S. & Soto, A. M. (2009). Bisphenol-Aand the great divide: a review of controversies in the field of endocrine disruption. *Endocrine Reviews*, *30*, 75–95.

Velez, M. P., Arbuckle, T. E. & Fraser, W. D. (2015). Female exposure to phenols and phthalates and time to pregnancy: the Maternal-Infant Research on Environmental Chemicals (MIREC) Study. *Fertility & Sterility*, *103*(4), 1011–1020.

Wang, J., Jenkins, S. & Lamartiniere, C. A. (2014). Cell proliferation and apoptosis in rat mammary gland following combinatorial exposure to bisphenol A and genistein. *BMC Cancer*, *14*, 379.

Wang, F., Zhang, Q., Zhang, X., Luo, S., Ye, D., Guo, Yi., Chen, S. & Huang, Y. (2014). Preeclampsia induced by cadmium in rats is related to abnormal local glucocorticoid synthesis in placenta. *Reproductive Biology Endocrinology*, *12*(1), 77.

Weinberger, B., Vetrano, A. M., Archer, F. E., Marcella, S. W., Buckley, B., Wartenberg, D., Robson, M. G., Kim, J., Azhar, S., Cavin, S., et al., (2014). Effects of maternal exposure to phthalates and bisphenol A during pregnancy on gestational age. *Maternal Fetal Neonatal Medicine*, *27*(4), 323-327.

Welshons, W. V., Nagel, S. C. & vomSaal, F. S. (2006). Large effects from small exposures. III. Endocrine mechanisms mediating effects of bisphenol A at levels of human exposure. *Endocrinology*, *147* (6), S56-S69.

Wens, B., De Boever, P., Verbeke, M., Hollanders, K. & Schoeters, G. (2013). Cultured human peripheral blood mononuclear cells alter gene expression when challenged with endocrine-disrupting chemicals. *Toxicology*, *303* (7), 17-24.

Wolff, M. S., Engel, S. M., Berkowitz, G. S., Ye, X., Silva, M. J., Zhu, C., Wetmur, J. & Calafat, A. M. (2008). Prenatal phenol and phthalate exposures and birth outcomes. *Environmental Health Perspectives*, *116*(8), 1092-1097).

World Health Organization (WHO) (2013). Mercury Exposure and Health Impacts among Individuals in the Artisanal and Small-Scale Gold Mining (ASGM) Community. Retrieved from http://www.who.int/ipcs/assessment/public_health/mercury_asgm.pdf.

World Health Organization (WHO), (2014). Childhood lead poisoning. Retrieved from http://www.who.int/ceh/publications/leadguidance.pdf.

World Health Organization (WHO), (2014). WHO recommendations on interventions to improve preterm birth outcomes. Retrieved from http://apps.who.int/iris/bitstream/10665/183037/1/9789241508988_eng.pdf.

Wu, K., Xu, X., Liu, J., Guo, Y., Li, Y. & Huo, X. (2010). Polybrominated diphenyl ethers inumbilical cord blood and relevant factors in neonates from Guiyu, China. *Environmental Science Technology*, *44*, 813-819.

Wu, Y., Chen, X., Zhou, Q., He, Q., Kang, J., Zheng, J., Wang, K., Duan, T. & Zhou, Y. (2014). ITE and TCDD differentially regulate the vascular remodeling of rat placenta via the activation of AhR. *PLoS One*, *9*(1), e86549.

Xu, X., Liu, J., Zeng, X., Lu, F., Chen, A. & Huo, X. (2014). Elevated serum Polybrominated Diphenyl Ethers and alteration of thyroid hormones in children from Guiyu, China. *PLoS One*, *9*(11), e113699.

Zhao, Y., Chen, L., Li, L. X., Xie, C. M., Li, D., Shi, H. J. & Zhang, Y. H. (2014). Gender-specific relationship between prenatal exposure to phthalates and intrauterine growth restriction. *Pediatric Research*, *76*(4), 401-408.

BIOGRAPHICAL SKETCH

Tandra R. Chakraborty

Science 113, Department of Biology Adelphi University One South Avenue, Garden City, New York 11530

Voice: 516-877-4202

e.mail: chakraborty@adelphi.edu

website: http://home.adelphi.edu/~ch17582

Education

2000-1993 Ph.D. in Biochemistry, University of Calcutta, India

1993-1991 MS. in Biochemistry, University of Calcutta, India

1991-1989 B.S. with Chemistry Honors, University of Calcutta, India

Professional Experience

Research & Teaching

Sept 2012 Associate Professor, Department of Biology, Adelphi University

2006-2012 Assistant Professor, Department of Biology, Adelphi University

Publications

Refereed Journal Articles ***undergraduate student, *Graduate student*

1. **Chakraborty, T.R.,** Alicea, E.**, Chakraborty, S. Relationship between urinary biomarkers of phytoestrogen, phthalates, phenols, and pubertal stages in girls. Adolescent health, Medicine and Therapeutics 2012 Jan 6; 3: 17-26
2. Ng, K.Y. **, Yong, J. ** and **Chakraborty, T.R.** Estrous Cycle in *ob/ob* and ovariectomized female mice and its relation with estrogen and leptin. 2010 Physiology & Behavior 99 (1) 12, 125-130.
3. Watson, E., Fargali, S., Okamoto, H., Sadahiro, M., Gordon, R.E., **Chakraborty, T.,** Sleeman, M.W. and Salton, S.R. Analysis of knockout mice suggests a role for VGF in the control of fat storage and energy expenditure. BMC Physiology 2009 28; 9(1): 19

4. Roy, J.R. [**], Chakraborty, S. and **Chakraborty, T.R**. Estrogen-like Endocrine Disrupting Chemicals Affecting Puberty in Humans- A Review Medical Science Monitor 2009 15(6): RA137-45.
5. Chakraborty, S., Sachdev, A[**]., Salton, S.R.J and **Chakraborty, T.R.** Expression of Estrogen Receptors in the Hypothalamic Arcuate Nucleus of *ob/ob* and Agouti Mice. Brain Research 2008 1089: 21-32
6. **Chakraborty, T.R**., Tkalych, O., Nanno, D., Garcia, A. L., Devi, L.A. and Salton, S.R.J. Regulation of VGF and PEN peptides in mouse brain in response to high fat or cold stress conditions. Brain Research 2006 17: 21-32
7. **Chakraborty, T.R.,** Rajendren, G.V. and Gore, A.C. Expression of the estrogen receptor alpha in the anteroventral periventricular nucleus (AVPV) of hypogonadal (hpg) mice. Experimental Biology & Medicine 2005 230(1): 49-56
8. **Chakraborty, T.R.,** and Gore, A.C. Aging-related changes in ovarian hormones, their receptors, and neuroendocrine function. Experimental Biology & Medicine 2004 229: 977-87
9. **Chakraborty, T.R.,** Hof, P., Ng, L.[**] and Gore, A.C. Stereological analysis of expression of estrogen receptor alpha in hypothalamus and its regulation by aging and estrogen. Journal of Comparative Neurology 2003 466(3): 409-421
10. **Chakraborty, T.R.,** Ng, L. [**] and Gore, A.C. Age-related changes in estrogen receptor beta in rat hypothalamus: a quantitative analysis. Endocrinology 2003 144(9): 4164-4171
11. Salama, J.[**], **Chakraborty, T.R.,** Ng, L.[**] and Gore, A.C. Effects of polychlorinated biphenyls (PCBs) on estrogen receptor □□expression in the antereoventral periventricular nucleus (AVPV). Environmental Health Perspectives 2003 111 (10): 1278-1282
12. **Chakraborty, T.R.,** Ng, L.[**] and Gore, A.C. Colocalization and hormone regulation of estrogen receptor alpha and NMDA receptor in the hypothalamus of female rats. Endocrinology 2003 144 (1): 299-305
13. Balija, V.S.[*], **Chakraborty, T.R.** and Haldar, D. Phospholipid synthesizing enzymes of the mitochondrial outer membrane: In search of their catalytic sites. Recent Research Developments in Lipids Research 2001 4: 1-11
14. Balija, V.S.[*], **Chakraborty, T.R.,** Nikonov, A., Morimoto, T. and Haldar, D. Identification of two transmembrane regions and a

cytosolic domain of rat mitochondrial glycerophosphate acyltransferase. Journal of Biological Chemistry 2000 275 (41): 31668-31673

15. **Chakraborty, T.R.,** Vancura, A., Balija, V.S.* and Haldar, D. Phosphatidic acid synthesis in mitochondria: Topography of formation and transmembrane migration. Journal of Biological Chemistry 1999 274(42): 29786-29790
16. **Roy, T.,** Chakraborty, S., Das, H., Chakraborty, B.N. and Bhattacharya, A.K. *In vitro* induction of nuclear DNA dispersion of ram spermatozoa. Journal Basic Applied Biomedicine 1995 3(3): 25-29

Non-Refereed Papers/Citations

17. Miller, A., **Wong, N., * Chakraborty, S., Adhikary, D. and **Chakraborty, T.R.** Differential Effect of High Fat Diet on Male Mice on Energy Intake, Hormones, Cellularity and Morphology of Tissue. American Society of Biochemistry and Molecular Biology, 2015
18. Wong, N., * Donthireddy L. * Babu, S. *, Adhikary, D. and Chakraborty S., **Chakraborty, T.R.** Differential Effect of High Fat Diet on Tissue M orphology and hormonal levels in male and female mice. American Society of Biochemistry and Molecular Biology, 2014
19. Donthireddy L., * Babu, S., * Wong, N., * Adhikary, D. and Chakraborty S., **Chakraborty, T.R.** The Adverse effects of High Fat Induced Obesity On Females Reproductive Cycle and Hormones. Delphi Research Conference 2013
20. Eldib, A., ** Eliscar, T., **Ng, K ** and **Chakraborty, T.R.** Effect Of Very High Fat Diet On Body Weight, Hormonal Levels And Estrous Cycle. American Society of Biochemistry and Molecular Biology, 2013
21. Eldib, A. **, Eliscar, T., ** Ng, K ** and **Chakraborty, T.R.** Effect Of Very High Fat Diet On Body Weight, Hormonal Levels And Estrous Cycle. American Society of Biochemistry and Molecular Biology, 2013

22. Eliscar, T. **, and **Chakraborty, T.R.** Effect Of Very High Fat Diet On Body Weight, Hormonal Levels And Estrous Cycle. CSTEP Conference, 2012
23. Alecea, E., Yohanan, D., Cohen, J.**, Addya, S., Chakraborty, S. and **Chakraborty, T.R.** Neuroprotection of estrogen on hypoglycemic attack is through AKT/GSK3 pathway. American Society of Biochemistry and Molecular Biology 2012 (Undergraduate Travel Award)
24. Alecea, E., and **Chakraborty, T.R.** Neuroprotection of estrogen on hypoglycemic attack is through AKT/GSK3 pathway. CSTEP Conference, 2012
25. **Chakraborty, T.R.** Effect of high fat diet on body weight, hormonal levels, morphology and estrous cycle. Indian Science Congress, 2013
26. Babu, S. and **Chakraborty T.R.** _'Effect of Very High Fat Diet on Hormone Levels and Body Weight in Female Mice,' *Metropolitan Association of College and University Biologists*, 2012
27. Reddy, L., and **Chakraborty T.R.** 'Effect of Very High Fat Diet on Hormone Levels and Body Weight in Male Mice,' *Metropolitan Association of College and University Biologists*, 2012
28. Ng, K**., Yong, J. **, and **Chakraborty, T.R.** Estrous Cycle in *ob/ob* and ovariectomized female mice and its relation with estrogen and leptin. Council On Undergraduate Research, Research highlights 5/14/2010

Invited Papers and Presentations (selected only)

Chakraborty, S., Babu, S.*, Adhikary, D., and **Chakraborty, T. R.** Long term exposure of Very High Fat Diet on Energy Intake, Hormones, Cellularity and Morphology of Tissue- a Male Rodent Study. American Society of Biochemistry and Molecular Biology, 2016

Chakraborty, T.R. Pros and Cons of Active Learning. American Society of Biochemistry and Molecular Biology, 2016

Miller, A.**, Wong, N.*, Chakraborty, S., Adhikary, D. and **Chakraborty, T.R.** Differential Effect of High Fat Diet on Male Mice on Energy Intake, Hormones, Cellularity and Morphology of Tissue. American Society of Biochemistry and Molecular Biology, 2015

Chakraborty, T.R., Babu S.*, Donthireddy, L.* 'Differential Effect of High Fat Diet in Men and Women,' Invited speaker at Indian Physiological Society, Dec, 2013

Chakraborty, T.R., Babu S.*, Donthireddy, L.* Effect of high fat diet on body weight, hormonal levels, morphology and estrous cycle. Invited speaker in the 100th Indian Science Congress, Jan, 2013

Alecea, E.**, Yohanan, D.**, Cohen, J.** Addya, S., Chakraborty, S. and **Chakraborty, T.R.** Neuroprotection of estrogen on hypoglycemic attack is through AKT/GSK3□ pathway. American Society of Biochemistry and Molecular Biology, 2012

Yohanan, D.**, Cohen, J.**, Min, I. *, Addya, S., Chakraborty, S. and **Chakraborty, T.R.** Neuroprotection by estrogen on hypoglycemic injury on hypothalamic cells at the cellular and molecular level. American Society of Biochemistry and Molecular Biology, 2011 *(Undergraduate Travel Award)*

Ng, K.Y.**, Yong, J.**, and **Chakraborty, T.R.** Effect of high fat diet on reproductive cycle. American Society of Biochemistry and Molecular Biology, 2011

Yong, J.P-W. **, Ng, K.Y. ** and **Chakraborty, T.R.** Are estrogen and leptin the sole factors affecting the estrous cycle of, *ob/ob* and ovariectomized mice? American Society of Biochemistry and Molecular Biology, 2010 ***undergraduate student (Undergraduate Travel Award)*

Ng, K.Y.**, Yong, J. **, and **Chakraborty, T.R** Effect of estrogen and leptin in the estrous cycle of *ob/ob* and ovariectomized mice? American Society of Biochemistry and Molecular Biology, 2009 ***undergraduate student (Undergraduate Travel Award)*

Chakraborty, T.R., Chu, P.*, and Chakraborty, S. Effect of atrazine on the hypothalamic cell Line. American Society of Biochemistry and Molecular Biology, 2009

Loera, M.*, and **Chakraborty, T.R.** Neuroprotective effect of estrogen on hypoglycemic injury in hypothalamic cells. American Society of Biochemistry and Molecular Biology, 2008 *(Minority Graduate Travel Award)*

Nikita, R.**, Sachdev, A.** and **Chakraborty, T.R.** Correlation of Estrous cycle with estrogen and leptin levels in *ob/ob* and Agouti mice. American Society of Biochemistry and Molecular Biology, 2008 *(Undergraduate Travel Award)*

Chakraborty, T.R., Loera, L.*, and Chakraborty, S. What is the pathway by which estrogen promotes the neuroprotective effect on hypoglycemic injury in hypothalamic cell lines. American Society of Biochemistry and Molecular Biology, 2008 *(Faculty Travel Award)*

Chakraborty, T.R., Loera, M.*, and Chakraborty, S. Neuronal protection from hypoglycemic injury by estrogen. Endocrine Society, 2008

Akhtar, A.** and **Chakraborty, T.R.** Effect of estrogen, estrogen receptor and neuropeptides on feeding and reproduction. Endocrine Society, 2007

Sachdev, A.,**, and **Chakraborty, T.R.** Cross talk of leptin and estrogen in the neuronal circuits to affect reproduction and feeding circuits. American Society of Biochemistry and Molecular Biology, 2007 *(Undergraduate Travel Award)*

Chakraborty, T.R., and Chakraborty, S. Estrogen receptors in obesity and their signaling pathways. American Society of Biochemistry and Molecular Biology, 2007

Reports published

1. Wong, N.,* Donthireddy L.* Babu, S.*, Adhikary, D. and Chakraborty S., **Chakraborty, T.R.** Differential Effect of High Fat Diet on Tissue Morphology and hormonal levels in male and female mice. ARC, 2013 *2nd prize in graduate poster presentation*
2. Babu, S. and **Chakraborty T.R.** ‘Effect of Very High Fat Diet on Hormone Levels and Body Weight in Female Mice,’ MACUB, 2012
3. Reddy, L., and **Chakraborty T.R.** ‘Effect of Very High Fat Diet on Hormone Levels and Body Weight in Male Mice,’ MACUB, 2012
4. Yong J.P-W. **, Ng K.Y. ** and **Chakraborty T.R.** Are estrogen and leptin the sole factors affecting the estrous cycle of, *ob/ob* and ovariectomized mice? The Annual Adelphi Research Conference, 2010
5. Yong J, Ng K, and **Chakraborty T.R.** Does estrogen and leptin effect the estrous cycle of *ob/ob* and ovariectomized mice? The Annual Adelphi Research Conference, 2009
6. Sachdev, A., and **Chakraborty, T.R.** Regulation of Feeding and Reproduction by Estrogen and Leptin. The Fourth Annual Adelphi Research Conference, 2007
7. Sachdev, A. and **Chakraborty, T.R.** Obesity and Reproduction, Biology Seminar Adelphi University Biology Department 2007
8. Akhtar, A., and **Chakraborty, T. R.** Effect of estrogen, estrogen receptor and neuropeptides on feeding and reproduction. The Fourth Annual Adelphi Research Conference, 2007

9. Akhtar, A., and **Chakraborty, T. R.** Regulation of Feeding Circuits by Hormones and Neuropeptides. Biology Seminar Adelphi University Biology Department 2006-2007
10. Loera, M., and **Chakraborty T.R.** Neuroprotective Effect of Estrogen on Hypothalamic injury in Hypothalamic cells. The Fourth Annual Adelphi Research Conference, 2007

Thesis Publication (Graduate and Undergraduates)

Non-alcoholic Fatty Liver and Alcoholic Fatty Liver, Biology Department, Adelphi University, 2016

Sleep Apnea Among Minorities, Biology Department, Adelphi University, 2016

DNMT (DNA Methyltransferase) And Diabetes Biology Department, Adelphi University, 2016

High Fat and Liver, Biology Department, Adelphi University, 2016

The compared effects of a Low Fat Diet and High Fat Diet on Female C57BL/6J mice, Biology Department, Adelphi University, 2015

Environmental Pollutants and Pregnancy, Biology Department, Adelphi University, 2015

Obesity and Metabolic Dysfunction, Biology Department, Adelphi University, 2013

Effect of Very high Fat diet on body weight, hormonal levels and estrous cycle in female mice, Biology Department, Adelphi University, 2013

Effects of very High Fat on body weight, hormonal levels and Estrous cycle in Mice, Biology Department, Adelphi University, 2013

Estrogen mediated Neuroprotection on N38 Mouse Hypothalamic Cells Under Hypoglycemic Conditions is through the AKT/GSK 3b pathway, Biology Department, Adelphi University, 2013

Neuroprotective Effects of Estrogen Against Hypoglycemic Injury on Hypothalamic Cells, Biology Department, Adelphi University, 2012

Effects of a High fat diet on Body weight, hormonal levels and the estrous Cycle, Biology Department, Adelphi University, 2012

Honors and Awards (selected only)

- Invited speaker at Adelphi Technology Conference, 'Strategies of Effective learning in Anatomy and Physiology Laboratory using iPad Applications' 2015

- Invited speaker at Adelphi Technology Conference, 'Pros and Cons of Active Learning' 2015
- Invited foreign delegate to centenary celebration of Indian Science Congress, 2013
- 'Faculty Development Career Grant,' Adelphi University, Jan 2011-Dec 2011
- Mentor of the student recipient for Undergraduate student Travel Award at ASBMB, 2011
- Nominated for 'Teaching Excellence Award,' 2010 at Adelphi University for non-tenured teaching award
- 'Individual Merit Award,' Adelphi University, 2009
- 'Faculty Development Career Grant,' Adelphi University, Jan 2008-Dec 2008
- American Society of Biochemistry and Molecular Biology, Undergraduate Faculty Travel ward, 2007 & 2008
- Mentor of the student recipient for Undergraduate student Travel Award at ASBMB, 2008
- Mentor of the gradute student recipient for Minority graduate Travel Award at ASBMB, 2007
- 'Frederick Bettelheim Research Award,' March 1st 2007- Feb 28th, 2008
- American Society of Biochemistry and Molecular Biology, Undergraduate Faculty Travel Award, 2007
- Mentor of the student recipient for Undergraduate student Travel Award at ASBMB, 2007
- 'Faculty Development Career Grant,' Adelphi University, Jan 2007-Dec 2007

Professional Achievements/Activities

I. Judge in Competitions

- Served as a panel judge in the Long Island Science and Engineering Fair, 2011 to 2015
- Served as a panel judge at Syosset Science fair 2012 to 2015
- Served as a panel judge at the Research Association Fair, 2014
- Served as a panel judge for undergraduate poster presentation at ASBMB, 2007, 2008 & 2010

- Judge for Junior Science and Humanities Symposium, Adelphi University, 2007

II. Invited Speaker

- Invited speaker in Queensboro Community College, Life Sciences, 2014
- -Invited speaker in the Indian Physiological Society, 2013
- Invited International delegate speaker in the Indian science Congress, Jan 2013
- Invited speak for the Honors Colloquium, Adelphi University, 20012

III. Service

Member of the editorial board of an open access, peer-reviewed journal titled 'ISRN Endocrinology'

Reviewer of 'Brain Research'

Reviewer of 'Annals of Anatomy'

Reviewer in the journal 'Journal of Neurocytochemistry'

Reviewer in the journal 'Experimental Biology & Medicine'

Reviewer in the journal 'Neurobiology of Aging'

Academic and Professional Societies

1998- present Member of 'American Society of Biochemistry and Molecular Biology'

2001- 2009 Member of 'Endocrine Society'

2000- 2007 Member of 'Society for Neuroscience'

2001- 2008 Member of 'Women in Endocrinology'

In: Endocrine Disrupting Chemicals
Editor: Sabrina Johnston
ISBN: 978-1-63485-231-9

Chapter 4

EXPOSING THE HIDDEN SOURCES OF ENDOCRINE-DISRUPTING CHEMICALS AND THEIR RELATED POTENTIAL EFFECTS ON HUMAN HEALTH

Karim N. Jallad*
American University of Kuwait,
College of Arts and Sciences, Safat, Kuwait

ABSTRACT

In the recent years, or to be exact since the middle of the last decade, the possible adverse effects of synthetic (man-made) and naturally occurring chemicals on the human and animal endocrine system have been the topic of many scientific and medical researches. These concerns were even highlighted further when the research began to show that these chemicals can influence the endocrine system at very low concentrations close to parts per trillion levels where the effects may be cumulative, potentially appearing in subsequent generations. Thus, these resulting effects might be both irreversible and threatening to the human sustainable development. These chemicals, known as Endocrine Disrupting Chemicals (EDCs), can be found in numerous places and products throughout the world since they are used, just to name a few, in

* Corresponding Author: Email: kjallad@auk.edu.kw (965) 2 224 8399 ext. 3720 (Voice), (965) 2 573 7039 (Fax).

turning on genes that will build a new protein. The hormone-receptor complex switches on or switches off specific biological processes in cells, tissues, and organs [2].

Although the amount of scientific information available for EDCs is continually increasing, the knowledge of the exact mechanisms and pathways of EDCs' actions are still limited. It is inferred from research that EDCs can act at multiple sites of the endocrine system through multiple mechanisms of action as follows; (1) they can mimic the biological activity of a hormone by binding to a cellular receptor, leading to an unwarranted or unneeded response by initiating the cell's normal response to the naturally secreted hormone at the wrong time or to an excessive amounts (agonistic effect); (2) they can bind to the receptor but not activate it, thus preventing the binding of the naturally produced hormone to the cell's receptor (antagonistic effect); (3) they can alter the amounts of hormones circulating in the blood by binding to the transport proteins present within the blood; (4) they can affect the synthesis or the breakdown rates of the natural hormones by interfering with the biological metabolic process in the body;(5) they can affect the cellular activity that involves regulating metabolic activity in cells, including cancer cells, as well as proliferation of cells and tumor progression by binging to estrogen related receptors; (6) they can interact directly with the molecules, thus bypassing the binding mechanism with endocrine receptors causing a cellular change that might contribute to the risk of developing cancer [3, 4, 5, 6]. In addition, EDCs may hamper the effects of the gender specific steroidal hormones on the brain, the pituitary gland, the gonads and the accessory sex organs, such as the uterus and the mammary glands in females and the prostate and seminal vesicles in males [7]. Such effects were supported by epidemiological studies where EDCs were shown to be associated with a variety of human health effects such as effects on reproduction, endometriosis, precocious puberty, neural function, immune function, and different types of cancer (thyroid, breast, prostate, testicular, and endometrial).

When it comes to international response regarding EDCs, scientific and public concerns regarding the potential adverse health effects resulting from a synergetic exposure to different EDCs at very low doses has grown. In 2002, the WHO International Program for Chemical Safety (IPCS) assessed the relationship between the EDCs and the adverse human health effects listed above, and concluded that EDCs play a major role in causing such effects even though the data regarding the dose-response relationship and exposure levels were scarce. Nevertheless, since then, scientific studies have demonstrated broadly and repeatedly the low dose effects of EDCs in many tissues and models [8, 9, 10, 11].

Currently, there are more than 87,000 chemicals produced and used worldwide, many of which have benefited society widely. Their number and diversity make identifying potential EDCs very difficult. Thus, identifying EDCs is the first step to reduce the potential health risks linked to these chemicals. A common international practice to control EDCs is to reduce their release in the environment, thus reducing their related potential health risks in humans. Over the past decades, the use of many toxic chemicals that have illustrated to cause endocrine disrupting effects was either banned or restricted [12, 13, 14]. There are no authoritative lists of either banned or restricted EDCs; however, all the environmental organizations (US Environmental Protection Agency-US EPA, WHO, ICPS, European Commission-EC, and The Organization for Economic Cooperation and Development-OECD) have taken initiatives to identify EDCs and research their related potential risk to humans' health and the environment. Fulfilling that initiative, these organizations classified the EDCs into three categories as follows; (1) Category 1 include chemicals that have illustrated to cause endocrine disrupting activities in at least one species using intact animals; (2) Category 2 include chemical that have shown endocrine disruption evidence in vitro; (3) Category 3 consist of chemicals with no data or evidence of any causation of endocrine disruption in any biological medium. Currently, The EC lists 194 EDCs as "Category 1" chemicals while the USEPA lists 134 [12, 15]. The listed chemicals, which are either high production or persistent ones, are mainly organochlorine pesticides or insecticides, industrial chemicals including by products such as dioxins and polychlorinated biphenyls (PCBs), bisphenol A (BPA), styrene, phthalates, organotins, and alkyl phenols including nonyl phenols.

It is worth noting that in 2001, more than 100 countries signed the Stockholm Convention (A) on persistent organic pollutants (POPs), which commits them to reduce or eliminate the production, use, and release of 12 chemicals known as the "dirty dozen" comprised of aldrin, chlordane, DDT, dieldrin, dioxins, endrin, furans, heptachlor, hexachlorobenzene, mirex, PCBs, and toxaphene. The "dirty dozen" fall within the same chemical classification of "Category 1" EDCs listed both the USEPA and the EC since aldrin, chlordane, DDT, dieldrin, endrin, heptachlor, hexachlorobenzene, mirex, and toxaphene are pesticides or insecticides while hexacholorbenzene and PCBs are industrial chemicals and dioxins, furans, hexachlorobenzene, and PCBs are also by products from combustion processes spanning from industrial to personal such as cigarettes' smoke and vehicular emissions. In 2009, the Stockholm convention added 9 chemicals to the original POP list. Five are pesticides namely; chlordecone, alpha hexachlorocyclohexane, beta hexachlorocyclohexane, lindane, and pentachlorobenzene, while four are industrial chemicals to include; hexabromobiphenyl; tetra-, penta-, hexa, and heptabromodiphenyl ethers; perfluorooctane sulfonic acids and its salts; and perfluorooctane sulfonyl fluoride.

In this chapter, we will mainly concentrate on "Category 1" anthropogenic EDCs that are either produced in high volumes or are considered to be POPs comprised of; organochlorine pesticides or insecticides, industrial chemicals including by products such as dioxins and PCBs, BPA, styrene, phthalates, organotins, and alkyl phenols. The sources of these chemicals will be revealed and their potential hazardous health effects on humans will be discussed.

Sources of Exposure to Endocrine Disrupting Chemicals

The wide range of chemicals shown to possess hormonal activity has been mentioned in the previous section and will be discussed profoundly under this one. Some EDCs are released into the environment intentionally such as insecticides, herbicides, and pesticides, but for the most environmental contaminants, release is unintentional. Unintentional release of chemicals can occur throughout part or all of the chemical's life cycle through manufacturing, use, and disposal [16]. Exposure can also take place through leakage from landfill areas and distribution via sewage sludge [17, 18, 19]. Consequently, exposure to EDCs can occur via air, water, soil, food, and consumer products. The air concentration of any chemical is dependent on its inherent and its relative volatility when in contact with water, vegetation, and soils. In addition, meteorological conditions such as temperature, wind speed, and humidity will influence the concentration and exposure assessment regionally and globally [16]. A variety of pesticides, industrial chemicals, and hormones have been detected in surface water, thus making it unsafe to be consumed by humans and terrestrial species [20]. Similarly, a number of potential EDCs have been detected in soil and sewage sludge in different parts of the world [21, 22, 23, 24] where especially farm animals may be exposed to contaminated soil through grazing, thus contributing to human exposure through this food chain pathway. Consequently, ingestion of both established and potential EDCs via food intake is generally considered the major exposure route for humans that may lead to, depending on species and quantities consumed, to bio- accumulation and bio-magnification [16]. Finally, for humans, a wide range of consumer products such as cleaning products, personal products, furniture, cosmetics, and garden chemicals provide pathways of exposure via inhalation, ingestion, and dermal contact. Of particular concern is the potential for phthalates ester exposure in young children orally by chewing on toys and teething rings [25]. The EDCs discussed in this section are high priority industrial chemicals according to international authorities since they heavily impact human heath due to both their persistence in the environment and their high volume production.

Organochlorine pesticides are chlorinated hydrocarbons used to eradicate a variety of insects. They were introduced in the 1940's and were widely used; however, they were phased out gradually due to their environmental persistence and potential adverse effects on human health. These pesticides categorized as POPs under the Stockholm convention [26] include but not limited to; chlordane, DDT, hexachlorobenzene, mirex, chlordecone, and toxaphene. Chlordane is a man-made chemical that was used as a pesticide in the United States from 1948 to 1988. From 1983 until 1988, chlordane's only approved use was to control termites in homes and before that, it was also used as a pesticide on agricultural crops, lawns, and gardens and as a fumigating agent [27]. Dichlorodiphenyltrichloroethane (DDT) is a pesticide that was once widely used to control insects on agricultural crops and insects that carry diseases like malaria and typhus, but is now used in only a few countries to control malaria; however, after 1972, the use of DDT was no longer permitted in the United States except in cases of a public health emergency. Hexachlorobenzene (HCB) was widely used as a pesticide to protect the seeds of onions, sorghum, wheat, and other grains against fungus until 1965 and was also used to make fireworks, ammunition, and synthetic rubber. Currently, there are no commercial uses of HCB in the United States. Mirex and chlordecone are two separate, but chemically similar, manufactured insecticides that do not occur naturally in the environment. Mirex and chlordecone have not been manufactured or used in the United States since 1978. Mirex was used to control fire ants, and as a flame retardant in plastics, rubber, paint, paper, and electrical goods from 1959 to 1972. Chlordecone was used as an insecticide on tobacco, ornamental shrubs, bananas, and citrus trees, and in ant and roach traps. Toxaphene was one of the most heavily used pesticides in the United States in the 1970s and early 1980s till all uses were banned by 1990. It was used primarily to control insect pests on cotton and

antifouling paints for boats, and as pesticides. Mono and di substituted organotins such as monomethyl tin (MMT), dimethyl tin (DMT), mono n-octyl tin (MOT), and di n-octyl tin (DOT) are in general used as stabilizers in PVC for food contact materials [40, 41]. Consequently, humans are mainly exposed to organotins through food ingestion, since migration will occur through food contact materials and soil contamination.

Nonylphenol (NP) is a widely used industrial nonionic surfactant belonging to the family of alkylphenols. The addition of ethoxy groups to NP produces nonylphenol ethoxylates (NPEs) which are also used in the manufacturing of industrial surfactants at a lower rate (80%) compared to NP. Besides being moderately persistent in the environment, NP is also believed to be ubiquitous since it is the predominant biodegradation product of NPEs [42]. NP and NPEs are used as nonionic surfactants in order to reduce the surface tension of liquids used in lubrication, defoaming agents, scouring fibers, emulsifiers, wetting and de-wetting agents, and dyes. In addition, NP is used to produce an antioxidant, tris(nonylphenol) phosphate, to stabilize plastics and rubber [43]. NP can enter the body through inhalation of air polluted with NP, through ingestion of contaminated water or food, or by dermal contact with NP containing consumer products [44].

Endocrine Disrupting Chemicals' Toxicity and Endocrine Disrupting Effects

When it comes to assessing exposure to EDCs in human, it is very important and essential to consider the following factors namely; level, timing, and duration of exposure relative to the developmental age of the organism [16]. Thus, age at which exposure occurs is very critical when it comes to assessing the potential effects of EDCs. It appears that the following development periods; in utero, neonatal, childhood, and puberty are the most critical and extremely susceptible to EDCs interferences. Furthermore, it was illustrated that exposure to EDCs during brain development can have permanent effects, while similar exposures to fully a differentiated and developed brain will have no noticeable effects [45].

Both chlordane and DDT were classified as moderately toxic (Oral LD_{50} = 50 – 500 mg/kg) for acute toxicity targeting liver in mice and affecting reproduction in experimental animals [46]. In addition, they were categorized as carcinogens to humans by the WHO International Agency for Research on

Cancer (IARC). HCB was classified by WHO as extremely hazardous (Oral LD_{50} = 0 – 50 mg/kg) for acute toxicity [48] and by IARC as a carcinogen [47]. It targets the liver, skin, bones, kidneys, thyroid, ovaries, and central nervous system in humans [49]; in addition, it causes death, developmental, endocrine, immunological, and neurological toxicity. Both chlordecone and mirex are similar, since their acute oral toxicity is moderate, they are both carcinogens [50], and they target thyroid, kidneys, and central nervous system in animals [51]. Finally among the organochlorine pesticides, toxaphene is considered both highly toxic via acute oral exposure targeting the nervous system and biological developmental in animals [52] and is a carcinogen to humans [53]. Few organochlorine pesticides have a negative adverse effect on the thyroid gland in humans, in particular DDT and HCB, where both produce anti-thyroid effects in pregnant women [54, 55, 56]. Furthermore, exposure to organochlorine pesticides interfere with neurodevelopment in humans, since studies have shown that in utero exposure is directly associated with diminished psychomotor and mental functions to include memory, attention span, and verbal skills in children [57, 58]. Similar neurobehavioral traits were exhibited by children born to mothers' residing next to agricultural regions contaminated with organochlorine pesticides [59]. In addition, organochlorine pesticides have demonstrated in vitro to be estrogen receptor agonists since many studies attempted inconclusively to link both breast and prostate cancers to these chemicals; nevertheless, it was established that they act as androgen, a sex hormone produced in males and females, receptor antagonist [54].

Long term environmental exposure to Dioxins and "dioxin like" PCBs can lead to toxic consequences affecting immunotoxicity, neurodevelopment, thyroid, and steroidal hormones [60], while these compound are not classified as carcinogenic to humans [30, 61]. Epidemiological research studies conducted in the United States, Japan, and China have shown that exposed to PCBs adversely affected children neurobehavioral development illustrated by lowering their IQ [54]. Additional studies, demonstrated that exposure to dioxins may be affecting the sperm quality in young males [62]; while occupational exposure to PCBs strongly correlates to prostate cancer mortalities among workers at capacitor manufacturing facilities in the Unites States [54].

BPA has very low acute toxicity and is not categorized as carcinogenic in humans. A number of research studies have inconclusively demonstrated the adverse health effects of being exposed to BPA. Such effects included, behavioral changes in girls (aggression and hyperactivity), sexual dysfunction in male adults, reduced semen quality in males, and increased incidences of diabetes [31, 32].

Styrene is considered to have a low acute toxicity and a possible carcinogen in humans [63]. Few epidemiological studies have shown that chronic exposure to styrene can lead to central nervous system depression and epithelial surface irritations, while others failed to demonstrate a strong correlation between exposure to styrene and increased abortion rates, elevated risks of developing leukemia, lymphoma, and hepatic dysfunction (liver disease) [63, 64]. Additional studies have illustrated that occupational exposure to styrene results in altering the levels of secreted thyroidal hormones [65].

Besides phthalates, DEHP, BBP, and DBP have low acute toxicity since they do not accumulate in the body and usually are metabolized and excreted instantaneously [66, 67] while DEHP is a possible carcinogen in humans [68, 69]. Nevertheless, concerns were raised in published studies regarding the effects of phthalates on liver, kidneys, lungs, blood clotting, and development of male reproductive tract in humans [70, 71, 72]. Additional studies, have suggested that prenatal exposure to phthalates can lead to developmental and behavioral abnormalities in humans [73, 74], and also have an inverse effect on the thyroid function, since DEHP can act as a thyroid receptor antagonist [75], while another showed that concentrations of the phthalates metabolites in the human body are associated with abdominal obesity and insulin resistance [76]. It is of significant importance, to mention, that studies illustrated good evidence between phthalate exposure during fetal life and shorter anogenital index (AGI), incompletely descended testes leading to a small scrotum and small penis size, and lower IQs due to attention deficit hyperactivity disorder (ADHD) symptoms.

When it comes to organotins, there is no human data regarding toxicity or carcinogenicity. TBT and TPT have median acute toxicity in animal studies and have shown to affect both the immune system and reproductive development, where female snails became masculinized and demonstrated imposex illustrated by the development of an additional male sex organ [41]. A number of studies have linked organotins exposure to obesity and diabetes in humans, where it was suggested that humans, who have been exposed to obesogenic chemicals, are prone to store larger quantities of fat resulting in a lifelong struggle to maintain a healthy weight especially if they tend to lead a sedentary lifestyle with unhealthy or poor eating habits [77, 78, 79].

Finally, similar to organotins, there is no human data regarding toxicity or carcinogenicity of NP. Np has shown to affect the immune system, reproductive system, and brain development in animal studies [43]. Furthermore, there is no available human data pertaining to the endocrine disrupting effects of NP. A single study showed that NP can exhibit estrogenic activities in in-vitro and in-vivo assays, where adverse effects may target the testes leading to decreased sperm production and increased uterine weight; nevertheless, relatively high dose levels of exposure are warranted before such effects are predicted to occur [42, 80].

CONCLUSION

This chapter discussed profoundly the endocrine disrupting effects and the toxicity of selected EDCs. In addition, the health effects resulting from exposure to these chemicals were examined according to the available literature. To effectively manage EDCs in the environment is difficult and complex, since they are diverse, widespread, persistent, and do not warrant remediation efforts. In addition, there are other confounding factors that are more likely to complicate the eradication or the reduction of EDCs' use such as; (i) shortfall of scientific agreement on what makes a chemical an EDC, (ii) lack of manufactures' disclosure information regarding the chemicals used and possible exposure to EDCs, (iii) shortage of scientific data on health effects, fate and transport, and on environmental distribution, (iv) enormous diversity of compounds that maybe EDCs, (v) widespread low-level environmental contamination that does not beseech remediation processes, and (vi) many potential EDCs are in high demand by consumers and substitutes may not exist or have been poorly studied, thus education the public is a necessity in order to reduce personal exposure to potential EDCs through informed product selection and use. In summary, evidence suggesting that human health is adversely and universally affected by EDCs remains weak. There is definitely corroborating data manifesting that humans are vulnerable to EDCs at high levels of exposure; however, the effects from long-term and low-level exposure at sensitive development stages early in life should be rigorously and thoroughly studied.

REFERENCES

[1] WHO IPCS. Global assessment of the state-of-the-science of endocrine disruptors. WHO/PCS/EDC. 2002 [cited 2015 December 29]. Available from URL: http://www.who.int/ceh/publications/endocrine/en/.

[2] US EPA. What is the endocrine system? Endocrine disruption. 2015 [cited 2015 December 29]. Available from URL: http://http://www.epa.gov/endocrine-disruption/what-endocrine-disruption.

[3] The Environment Directorate-General of the European Commission. Endocrine disrupters. 2012 [cited 2016 January 02]. Available from URL: http://ec.europa.eu/environment/endocrine.

[4] Takayanagi, S; Tokunaga, T. Endocrine disruptor bisphenol A strongly binds to human estrogen-related receptor gamma (ERRgamma) with high constitutive activity. *Toxicol. Lett.*, 2006 167 (2), 95–105.

[5] De Coster, S; van Larebeke, N. Endocrine-Disrupting Chemicals: Associated Disorders and Mechanisms of Action. *J. Env. Pub. Health*, 2012 Article ID: 713696, 52 pages. from URL: http://www.hindawi.com/journals/jeph/2012/713696/.

[6] Bianco, S; Sailland; J. ERRs and cancers: effects on metabolism and on proliferation and migration capacities. *J. Steroid Biochem. Mol. Biol.*, 2012 130, 180–5.

[7] Wuttke, W; Jarry, H. Definition, classification and mechanism of action of endocrine disrupting chemicals. *Hormones*, 2010 9(1), 209–215.

[8] Vandenberg, LN; Colborn, T. Hormones and endocrine-disrupting chemicals: low-dose effects and nonmonotonic dose responses. *Endocr. Rev.*, 2012 33(3), 378–455.

[9] Diamanti-Kandarakis, E; Bourguignon, JP. Endocrine-disrupting chemicals: an Endocrine Society scientific statement. *Endocr. Rev.*, 2009 30(4), 293–342.

[10] Welshons, WV; Thayer, KA. Large effects from small exposures. I. Mechanisms for endocrine-disrupting chemicals with estrogenic activity. *Environ. Health Perspect.*, 2003 111(8), 994–1006.

[11] Hofercamp, L. Current use pesticides in Arctic media: 2000-2007. *Sci. Total Env.*, 2010 408, 2985–2994.

[12] Government of Canada. Order amending schedule I to the hazardous products act (bisphenol A), 2010 144(7). Available from: URL: http://gazette.gc.ca/rp-pr/p2/2010/2010-03-31/html/sor-dors53-eng.html.

[13] European Commission. Directives. Commission Directive 2011/8/EU of 28 January 2011. Amending Directive 2002/72/EC as regards the

restriction of use of Bisphenol A in plastic infant feeding bottles. *Official Journal of the European Union*. 2011. Available from: URL:http://eur-lex.europa.eu/LexUriServ/LexUriServ.do?uri=OJ:L:2011:026:0011:0014:EN:PDF.

[14] Australia Government, Department of Health and Aging NICNAS. Endocrine disrupting chemicals (EDCs). 2015 [cited 2016 February 5]. Available from URL: https://www.nicnas.gov.au/communications/publications/the-nicnas-bulletin/the-nicnas-bulletin5-may-2015/chemicals-that-alter-endocrine-hormone-system-functions.

[15] US EPA. Endocrine Disruptor Screening Program (EDSP). 2012 [cited 2016 February 5]. Available from URL: http://www.epa.gov/endocrine-disruption/endocrine-disruptor-screening-program-edsp-21st-century.

[16] WHO IPCS (2000). Human exposure assessment. Environmental Health Criteria 214. International Program on Chemical Safety. Geneva, Switzerland.

[17] Fara, G. M. (1999). The ICMESA accident: First intervention for the protection of man and environment. In: A. Ballarin-Denti, P.A. Bertazzi, S. Facchetti, R. Fanelli, & P. Mocarelli (Eds.), *Chemistry, Man and Environment* (pp. 3-16). Amsterdam: Elsevier Science.

[18] Tobin, P. (1986). Known potential sources of hexachlorobenzene. In: C. R. Morris, & J. R. P. Cabral (Eds.), *IARC, Hexachlorobenzene: Proceedings of an International Symposium*. Lyon, France: Oxford University Press.

[19] Daughton, CG; Ternes, TA. (1999) Pharmaceuticals and personal care products in the environment: Agents of subtle change? *Env. Health Persp.*, 1999 107(Suppl 6), 907-938.

[20] Mackay, D; Fraser, A. Bioaccumulation of persistent organic chemicals: Mechanisms and models. *Env. Pol.*, 2000 110(3), 375-391.

[21] Lega, L; Ladwig, G. Quantitative determination of organic priority pollutants in sewage sludge by GC/MS. *Chemosphere*, 1997 34, 1705-1712.

[22] Hale, RC; La Guardia, MJ. Persistent pollutants in land-applied sludges. *Nature*, 2001 412, 140-141.

[23] Kocan, A; Petrik J. Environmental contamination with polychlorinated biphenyls in the area of former manufacture in Slovakia. *Chemosphere*, 2001 43, 595-600.

[24] Stevens, J; Green, NJL. Survey of PCDD/Fs and non-ortho PCBs in UK sewage sludges. *Chemosphere*, 2001 44, 1455-1462.

[25] Steiner, I; Kubesch, K. Migration of phthalates from PVC-consumption goods. *Lebensmittel chemie,* 1999 53, 147-148.

[26] UN Environment Program. Stockholm Convention on Persistent Organic Pollutants (POPs). Geneva, Secretariat of the Stockholm Convention. 2011 [cited 2016 February 5]. Available from URL: http://chm.pops.int.

[27] Agency for Toxic Substances and Disease Registry (ATSDR). Draft Toxicological Profile for Toxaphene. U.S. Department of Health and Human Services. Public Health Service. 2010. Available from URL: http://www.atsdr.cdc.gov/toxprofiles/tp94.pdf.

[28] Center for Disease Control (CDC). Forth national report on human exposure to environmental chemicals. Organo- chlorine pesticides. 2012 [cited 2016 February 5] Available from URL: http://www.cdc.gov/exposurereport/pdf/fourthreport.pdf.

[29] WHO. Safety evaluation on certain food additives and contaminants: Polychlorinated dibenzodioxins, polychlorinated dibenzofurans, and coplanar polychlorinatedbiphenyls; *WHO Food Additives Series*: 48. Geneva: WHO; 2002. Available from URL: http://www.inchem.org/documents/jecfa/jecmono/v48je20.

[30] IARC. Summaries & evaluations: Polychlorinated dibenzo-para-dioxins. IARC Monographs on the Evaluation of Carcinogenic Risks to Humans, 1997 69, p. 33. Available from URL: http://www.inchem.org/documents/iarc/vol69/dioxin.html.

[31] EFSA. Scientific opinion on Bisphenol A: evaluation of a study investigating its neurodevelopmental toxicity, review of recent scientific literature on its toxicity and advice on the Danish risk assessment of Bisphenol A. *The EFSA Journal*, 2010 8(9), 1829. Available from: URL:http://www.efsa.europa.eu/en/efsajournal/pub/1829.

[32] Food and Agricultural Organization of United Nations (FAO UN) WHO. Toxicological and health aspects of Bisphenol A. Report of Joint FAO/WHO Expert Meeting November 2- 5 2010 and Report of stakeholder meeting on Bisphenol A. 2010. Ottawa, Canada 2011. Available from: URL:http://apps.who.int/iris/bitstream/10665/44624/1/97892141564274_eng.pdf.

[33] WHO Europe. Air quality guidelines, 2nd Edition. *Chapter 5.12 – Styrene*, 2010. [cited 2012 April 13]. Available from URL: http://www.euro.who.int/__data/assets/pdf_file/0018/123066/AQG2ndEd_5_12Styrene.pdf.

[34] WHO. Environmental health criteria 26. Styrene. 1983. Available from URL: http://www.inchem.org/documents/ehc/ehc/ehc26.htm.

[35] EFSA. Opinion of the scientific panel on food additives, flavorings, processing aids and materials in contact with food (AFC) on a request from the commission related to Bis (2- ethyl hexyl) phthalate (DEHP) for use in food contact materials. Question N° EFSA-Q- 2003-191. *The EFSA Journal*, 2005 243, 1-20. Available from URL: http://www.efsa.europa.eu/en/efsajournal/pub/243.

[36] Leadbitter J. Packaging materials 5. Polyvinyl Chloride (PVC) for food packaging applications. International Life Sciences Institute. June 2003. Available from URL: http://www.ilsi.org/Europe/Publications/R2003Pac_Mat5.pdf.

[37] US EPA. Technology transfer Network. Bis(2-ethylhexyl) phthalate (DEHP). 117-81-7. January 2000. Available from URL: http://www3.epa.gov/airtoxics/hlthef/eth-phth.html.

[38] ATSDR. Toxicology profile of Di(2-ethylhexyl)phthalate. Atlanta, GA: U.S. Department of Health and Human Services, Public Health Service. 2002. Available from URL: http://www.atsdr.cdc.gov/ToxProfiles/TP.asp?id=684&tid=65.

[39] WHO. Environmental Health Criteria 15. Tin and organotin compounds. 1980. Available from URL: http://www.inchem.org/documents/ehc/ehc/ehc015.htm.

[40] Technical University of Denmark. DTU Food. National Food Institute. Organotin compounds in food contact materials. 2010. [cited 2016 February 05]. Available from URL: http://www.food.dtu.dk/Default.aspx?ID=20917.

[41] EFSA. Opinion of the scientific panel on contaminants in the food chain on a request from the Commission to assess the health risks to consumers associated with exposure to organotins in foodstuffs (Question N° EFSA-Q-2003-110). *The EFSA Journal,* 2004 102, 1-119. Available from URL: http://www.efsa.europa.eu/en/efsajournal/doc/102.pdf.

[42] California Environmental Protection Agency. Toxicology profile for nonlyphenol 2009. Available from URL: http://www.opc.ca.gov/webmaster/ftp/project_pages/MarineDebris_OEHHA_ToxProfiles/Nonylphenol%20Final.pdf.

[43] Environmental Health Department Ministry of the Environment Government of Japan. Report on the test results of endocrine disrupting effects of nonylphenol on fish (Draft). 2001. Available from URL: http://www.oecd.org/dataoecd/7/4/2460701.pdf.

[44] European Union Joint Research Center (EU JRC). 4-nonylphenol (branched) and nonylphenol CAS Nos: 84852-15-3 and 5425154- 52-3 EINECS Nos: 284-325-5 and 246-672-0. Summary risk assessment. 2002. Available from URL: http://echa.europa.eu/documents/10162/6c460d8a-9f18-475f-823c-b8941e18fa3a.

[45] Jacobson, JL; Jacobson SW. Postnatal exposure to PCBs and childhood development. *Lancet*, 2001 358(9293), 1568-1569.

[46] WHO. Chlordane. Pesticide residues in food: evaluations part II toxicology. Geneva: WHO; 1986. Available from URL: http://www.inchem.org/documents/jmpr/jmpmono/v86pr03.htm.

[47] IARC. IARC monographs on the evaluation of carcinogenic risks to humans volume 53. Occupational exposures in insecticide application, and some pesticides-summary of data reported and evaluation. France: IARC; last updated 1999. Available from URL: http://monographs.iarc.fr/ENG/Monographs/vol53/mono53.pdf.

[48] WHO. The WHO recommended classification of pesticides by hazard and guidelines to classification 2009. Available from URL: http://apps.who.int/iris/bitstream/10665/44271/1/9789241547963_eng.pdf?ua=1.

[49] Toxicological Profile for Hexachlorobenzene. U.S. Department of Health and Human Services. Public Health Service. Agency for Toxic Substances and Disease Registry. 2002. Available from URL: http://www.atsdr.cdc.gov/ToxProfiles/tp90.pdf.

[50] IARC. IARC monographs on the evaluation of carcinogenic risks to humans volume 20. Some halogenated hydrocarbons. Summary of data reported and evaluation. France: IARC; last updated1999. Available from URL: http://monographs.iarc.fr/ENG/Monographs/vol1-42/mono20.pdf.

[51] Toxicological profile for mirex and chlodccone. U.S. Department of Health and Human Services. Public Health Service. Agency for toxic substances and disease registry. 1995. Available from URL: http://www.atsdr.cdc.gov/toxprofiles/tp66.pdf.

[52] ATSDR. Draft Toxicological Profile for Toxaphene. U.S. Department of Health and Human Services. Public Health Service. Agency for Toxic Substances and Disease Registry. 2010. Available from URL: http://www.atsdr.cdc.gov/toxprofiles/tp94.pdf.

[53] IARC. IARC Monographs on the Evaluation of Carcinogenic Risks to Humans Volume 79. Some Thyrotropic Agents Summary of Data Reported and Evaluation. France: IARC; last updated 2001. Available

from URL: http://monographs.iarc.fr/ENG/Monographs/vol79/mono79.pdf.
[54] Evans, R; Kortenkamp, A. State of the Art Assessment of Endocrine Disrupters. 2nd Interim Report. Project Contract number 070307/2009/550687/SER/D3. Part 1. Summary of the State of the Science. 2011. EC Environment. Available from URL: http://ec.europa.eu/environment/chemicals/endocrine/pdf/annex1_summary_state_of_scie nce.pdf.
[55] Chevrier, J; Eskenazi, B. Effects of exposure to polychlorinated biphenyl and organochlorine pesticides on thyroid function during pregnancy. *Am. J. Epid.,* 2008 168(3), 298-310.
[56] Roman GC. Autism: Transient in utero hypothyroxinemia related to maternal flavonoid ingestion during pregnancy and to other environmental antithyroid agents. *J. Neuro. Sci.*, 2007 62, 15-26.
[57] Marks, AR; Harley, K. Organophosphate pesticides exposure and attention in young Mexican-American children: the CHAMACOS Study. *Env. Health Persp.,* 2010 118, 1768-1774.
[58] Ribas-Fito, N; Torrent, M. In utero exposure to background concentrations of DDT and cognitive functioning among preschoolers. *Am. J. Epid.,* 2006 164, 955-962.
[59] Robert, EM; English, PB. Maternal residence near agricultural pesticide applications and autism spectum disorders among children in the California Central Valley. *Env. Health Persp.,* 2007 115, 1482-1489.
[60] WHO. Exposure to dioxins and dioxin-like substances: A major public health concern. WHO 2010. Available from URL: http://www.who.int/entity/ipcs/features/dioxins.pdf.
[61] WHO IARC. Monograph working group. Special Report: Policy – A review of human carcinogens – Part F: Chemical agents and related occupations. *The Lancet Oncology*, 2009 10(12), 1143-1144.
[62] Mocarelli, P; Gerthoux, PM. Dioxin exposure, from infancy through puberty, produces endocrine disruption and affects human semen quality. *Env. Health Persp.,* 2008 116(1), 73-77.
[63] WHO. Toxicological evaluation of certain food additives and food contaminants. WHO Food Additives Series no.19. 1982. Available from URL: http://www.inchem.org/documents/jecfa/jecmono/v19je15.htm.
[64] IARC. Summaries & evaluations: styrene, volume 82, 2002. Available from URL: http://www.inchem.org/documents/iarc/vol82/82-07.html.
[65] Ministry of Health and Welfare of Japan. Interim report Effects of Endocrine Disruptors on Human Health. 1998 Available from URL:

http://www.ffcr.or.jp/zaidan/ffcrhome.nsf/7bd44c20b0dc562649256502001b65e9/5e69e178bbad89414925681d001fa22d/$file/appendix-2.pdf.
[66] European Union Joint Research Center (EU JRC). Bis (2-ethylhexyl) phthalate (DEHP) CAS No: 117-81-7 EINECS No: 204-211-0 RISK ASSESSMENT. 2008. Available from URL: http://publications.jrc.ec.europa.eu/repository/bitstream/111111111/5648/1/dehpreport042.pdf.
[67] EU JRC. Union risk assessment report, benzyl butyl phthalate (BBP). CAS Nos: 85-68-7. EINECS Nos: 201-622-7. Risk Assessment. 2003. Available from URL: http://echa.europa.eu/documents/10162/060d4981-4dfb-4e40-8c69-6320c9debb01.
[68] IARC. Some chemicals in industrial and consumer products, food contaminants and flavorings, and water chlorination by-products. Vol 101. April 2011. Available from URL: http://monographs.iarc.fr/ENG/Meetings/index1.php.
[69] IARC. Summaries & evaluations butyl benzyl phthalate (Group 3). 1999. Available from URL: http://www.inchem.org/documents/iarc/vol73/73-04.html.
[70] EU JRC. Union Risk Assessment Report, Dibutylphthalte (DBP) EINECS No: 201-557-4. Risk assessment. 2003. Available from URL: http://publications.jrc.ec.europa.eu/repository/bitstream/111111111/5681/1/EUR%2019840%20EN.pdf.
[71] EU JRC. 1,2-benzenedicarboxylic acid, di-C8-10-branched alkyl esters, C9-rich and di- "isononyl" phthalate (DINP) CAS No: 68515-48-0, 28553-12-0 EINECS No: 271-090-9, 249-079-5. Risk assessment. 2003. Available from URL: http://publications.jrc.ec.europa.eu/repository/bitstream/111111111/5395/1/EUR%20207 84%20EN.pdf.
[72] EU JRC. 1,2-benzenedicarboxylic acid, di-C9-11-branched alkyl esters, C10-rich and di- "isodecyl" phthalate (DIDP). CAS Nos: 68515-49-1 and 26761-40-0. EINECS Nos: 271- 091-4 and 247-977-1. Risk assessment. 2003. Available from URL: http://publications.jrc.ec.europa.eu/repository/bitstream/111111111/5459/1/EUR%2020785%20EN.pdf.
[73] Engel, SM; Zhu, C. Prenatal phthalate exposure and performance on the Neonatal Behavioral Assessment Scale in a multiethnic birth cohort. *Neutotoxicology,* 2009 30, 522-528.
[74] Miodovnik, A; Engel, SM. Endocrine disruptors and childhood social impairment. *Neurotoxicology*, 2011 32, 261-267.
[75] Meeker JD and Ferguson KK. Relationship between urinary phthalate and bisphenol A concentrations and serum thyroid measures in U.S.

adults and adolescents from NHANES 2007-08. *Env. Health Persp.,* 2011; 119: 1396-1402.
[76] Stahlhut, RW; van Wijngaarden E. Concentrations of urinary phthalate metabolites are associated with increased waist circumference and insulin resistance in adult U.S. males. *Env. Health Persp.,* 2007 115(6), 876-882.
[77] Muncke L. Endocrine disrupting chemicals and other substances of concern in food contact materials: an updated review of exposure, effect and risk assessment. J. Steroid and Biochem. Mol. Bio., 2011 127(1-2), 118-127.
[78] National Toxicology Program The National Institute of Environmental Health Sciences (NTP NIEHS). NIEHS environmental factor article: NTP workshop investigates links between chemicals and obesity Organotins and Phthalates (version updated January 5, 2011). Available from URL: http://ntp.niehs.nih.gov/ntp/ohat/diabetesobesity/Wkshp/ OrgantinsPhthalates DraftLiteratureReviewV3formatted.pdf.
[79] Janesick, A; Blumberg, B. The role of environmental obseogens in the obesity epidemic. *Endocrine Update,* 2011 30(4), 383-399.
[80] WHO IPCS. Integrated Risk Assessment: Nonylphenol. Case study report prepared for the integrated risk assessment: Nonylphenol. WHO UNEP ILO International Program on Chemical Safety. 2004. Available from URL: http://www.who.int/ipcs/methods/Nonylphenol.pdf.

INDEX

#

A

B

C

D

E

F

G

H

I

J

K

L

M

N

O

P

Q

R

S

T

U

V

W